TOXINS AND TARGETS

TOXINS AND TARGETS

Effects of Natural and Synthetic Poisons on Living Cells and Fragile Ecosystems

Edited by

Dianne Watters
Martin Lavin
Queensland Cancer Fund Research Unit
Queensland Institute of Medical Research
Brisbane, Australia

David Maguire
Division of Science and Technology, Griffith University
Brisbane, Australia

John Pearn
Faculty of Medicine and Department of Child Health University of Queensland, Royal Children's Hospital Brisbane, Australia

Routledge
Taylor & Francis Group
LONDON AND NEW YORK

First published 1922 by Harwood Academic Publishers GmbH.
Published 2022 by Routledge

2 Park Square, Milton Park, Abingdon, Oxon OX14 4RN
605 Third Avenue Avenue, New York, NY 10017

Routledge is an imprint of the Taylor & Francis Group, an informa business

ISBN-13: 978-3-718-65194-8 [hbk]
ISBN-13: 978-1-315-07691-1 [ebk]

Library of Congress Cataloging-in-Publication Data

Toxins and targets / edited by Dianne Watters ... [et al.].

p. cm.
Includes bibliographical references and index.
ISBN 3-7186-5194-7
1. Toxins. 2. Toxicology. 3. Poisonous animals. I. Watters, Dianne, 1952–
[DNLM: 1. Toxins—metabolism. QW 630 T7554]
QP631.T68 1992
574.2' 1—dc20
DNLM/DLC
for Library of Congress 91-20865
CIP

Front cover photograph:
Morphological change caused by toxins: distortion (blebbing) of an hepatocyte after exposure to *Microcystis* toxin (refer to Chapter 6).

CONTENTS

PART II
DAMAGE TO ANIMAL CELLS

PART III
POLLUTION, PEOPLE AND PREVENTION

PREFACE

The last decade of the twentieth century has seen an unprecedented advocacy for the preservation of a safe environment, and the prevention of ill health from toxic chemicals. No scientist can ignore these issues. All have a responsibility to maintain a technical expertise in their specialist fields of endeavour. All have an inescapable ethical responsibility to promote an environment in which ecosystems remain balanced.

Ecotoxicology has become one of the most important interdisciplinary fields of study; coordination and cooperation among the many fields of biological and physical sciences are now among the most pressing concerns for contemporary society.

Decision makers whose roles include the promotion and control of the environment must look to lessons that can be learnt from biosystems that include animals other than man. We hope these pages may be of help to all who have a professional responsibility to maintain our world as a place of enduring beauty.

Dianne Watters
Martin Lavin
David Maguire
John Pearn

ACKNOWLEDGEMENTS

This book has been compiled with the help and encouragement of many people with a manifest commitment to promoting a better world environment. The contributors assembled at Mount Tamborine, Australia, in September 1989 for a major symposium and workshop sponsored by the Australian Biochemical Society and the International Union of Biochemistry. This publication is based on selected and updated reports from that conference. The manuscript was prepared with the dedicated assistance of Mrs Lisa Cooper, Ms Annabel Parkinson and Mrs Peggy Carter, of the University of Queensland. We also acknowledge the contributions of Professor C. J. Hawkins, Dr M. Hegarty, Dr E. Bennett, Dr R. Brinkworth, Dr J. de Jersey and the numerous postgraduate students who assisted in the organization of the conference.

INTRODUCTION

The biological world is an amazing collection of self-replicating organisms. Each organism is the end product of a series of evolutionary adventures. Each species has tested the environment and challenged other inhabitants of the biosphere in its competition for space, nutrients and reproductive success.

The chapters in this book detail the extent to which organisms from diverse habitats in the animal and plant kingdoms have adapted to handle toxins safely. Such a process has included the evolution of defence and prey-hunting activities on the one hand, and survival in an increasingly toxic environment on the other.

Recent activities of humans have accelerated the rate of introduction of new toxins into the environment, thereby contributing to the demise of many species. In order to control competitors for our food sources, we have assembled a large armament of pesticides and permitted their widespread and largely uncontrolled application.

It is not only in our agricultural pursuits that we apply biocides, for they have also become valuable tools in the fight against disease. Should we really be surprised that ill-considered assaults, using relatively non-specific toxins, produce adverse effects in the environment? When our own health or that of our progeny begins to suffer from the effects of these chemicals, it is certainly time to question the efficacy of such control measures.

We have become so dependent on the use of chemical control measures in agriculture and in veterinary and medical practice that it is unlikely we can reverse the trend without even more serious implications. We must therefore seek every means of promoting safer ways to produce, package and deliver toxins with improved specificity. Can we heed the valuable lessons woven into the fabric of evolution in order to achieve this goal?

The first section of this book is focussed on the underwater environment. It highlights the means by which marine organisms have adapted their toxins to be effective in a world of high aqueous solubility and infinite dilution. For survival in a watery habitat, sessile organisms are particularly dependent upon toxins for self defence, and several chapters are devoted to the types of toxins produced by such species. Three chapters focus on various aspects of coneshells. During evolution, these fascinating molluscs have acquired elaborate mechanisms for the synthesis, storage and delivery of venom to ensure success in prey-hunting and defence.

Part II is concerned mainly with venomous terrestrial organisms and the mechanisms by which they damage animal cells. What special features of their habitat constrain the evolution of different toxins and varying forms of venom delivery? Widely different life forms within the terrestrial environment are examined— animals and plants, bacteria and viruses. Without toxins to protect them, terrestrial sessile organisms such as cycads would be extremely vulnerable to predators. As with their sessile marine counterparts (e.g. sea-squirts and the soft corals), toxins are distributed throughout their tissues. Despite their mobility, spiders and snakes nevertheless exhibit the need for highly specialized venom production and delivery mechanisms. Such predator-prey relationships may be extremely specialized, and therefore are ecologically fragile.

In the last section we consider the way in which human society attempts to investigate toxicity, and to control poisons, be they synthetic or naturally occurring ones. The first step in this process is to alert ourselves to the insidious nature of toxicity. Three chapters in this section encompass this theme. Another describes how a government agency might tackle some of the difficult problems associated with monitoring toxicity, and how it might control the use of toxic material.

Many of the chapters in this book allude to the interaction between toxin and receptor— the molecular target. As our knowledge of natural toxin structures and receptor topology increases, our ability to describe this interaction also improves. Currently there is an increasing awareness and appreciation that toxin interactions depend upon the subtleties of spatial arrangements. This exciting theme is developed in several chapters in this book.

The development and preservation of safe environments will depend on greater factual knowledge and a quest for novel approaches to solving the escalating problems. It is hoped that this book will be a step towards this end.

CONTRIBUTORS

This section contains details of editors and senior authors only.

Dr Nor Azila is a lecturer in biochemistry within the Faculty of Medicine of the University of Malaya, Kuala Lumpur, Malaysia. Her doctorate was awarded for studies of the metabolism of phosphoinositides in the adrenal medulla. Her major current research interests include biochemical studies of marine toxins, especially those found in jellyfish. (Department of Biochemistry, Faculty of Medicine, University of Malaya, Kuala Lumpur, Malaysia, 59100.)

Dr Bruce F. Bowden is a lecturer in organic chemistry at James Cook University of North Queensland, Townsville, Australia. His major research interest is in the field of marine chemistry, with particular emphasis on the organic chemistry of soft corals. (Department of Chemistry, James Cook University of North Queensland, Townsville, Australia, 4811.)

Dr Ross I. Brinkworth has held research positions at Temple University (Philadelphia), Case Western Reserve University (Cleveland), Monash University and the Victorian College of Pharmacy. His research interests have included enzyme inhibitors and receptor binding studies. In his current appointment as assistant professor of biochemistry at Bond University, he is working on AIDS research. (Centre for Drug Design and Development, Bond University Research Park, Gold Coast, Australia, 4217.)

Dr Michael F. Capra is director of the Centre for Biological Population Management, School of Life Sciences, Queensland University of Technology. His research interests cover marine toxins, localization of heavy metals in marine organisms and mariculture. (Queensland University of Technology, 2 George Street, Brisbane, Australia, 4000.)

Dr Des W. Connell is director of the Queensland Government Chemical Laboratory and Associate Professor in the Division of Australian Environmental Studies at Griffith University. His principal research interests include the distribution and behaviour of toxicants in the environment. (Government Chemical Laboratory, Coopers Plains, Queensland, Australia, 4108.)

Mr Michael Dobbie is a doctoral student under the supervision of Associate Professor Alan Jones at the University of Sydney. Their research is concerned with the metabolism of normal spermatozoa and the action of male antifertility compounds using pig spermatozoa as a model system. (Department of Biochemistry, University of Sydney, Australia, 2006.)

Associate Professor Peter R. Dunkley is head of medical biochemistry at the University of Newcastle. His research is focussed on the role of second messengers and protein phosphorylation in the secretion of hormones and neurotransmitters. (University of Newcastle, Newcastle, Australia, 2308.)

Professor Ian R. Falconer is dean of the Faculty of the Sciences at the University of New England. His research interests include cyanobacterial toxins, hormone action and cell and tissue control. (Department of Biochemistry, Microbiology and Nutrition, University of New England, Armidale, Australia, 2351.)

Dr David R. Hillyard trained as a pathologist. In his present position in the Department of Pathology at the University of Utah, he works with the genes involved in the synthesis of conotoxins. (Department of Pathology, University of Utah, Salt Lake City, Utah 84112, USA.)

Dr Grahame J. Kelly is a lecturer in the Department of Biology at the Queensland University of Technology. He is a plant biochemist working on photosynthesis, particularly photosynthetic carbon metabolism. A major current interest is to determine the interactions between photosynthesis and the greenhouse effect.

(Queensland University of Technology, 2 George Street, Brisbane, Australia, 4000.)

Professor Martin F. Lavin is senior principal research fellow and head of the Division of Cancer Research at the Queensland Institute of Medical Research. He has made major contributions to understanding the biochemical defect in the human genetic disease, Ataxia telangiectasia. Included among his many research interests are cytotoxic compounds and novel proteins from ascidians, and the embryological development of ascidians as a model for vertebrate development.(Queensland Cancer Fund Research Unit, Queensland Institute of Medical Research, Herston Rd, Brisbane, Australia, 4029.)

Dr David Maguire is a lecturer in the Division of Science and Technology, Griffith University. A long period of domiciliary bliss in North Queensland exposed him to the beauty and the dangers of the creatures of the Great Barrier Reef. During this time he began to focus on the biosynthesis of *Conus* toxins. He divides his research time between these studies and the development of medical diagnostics. (Division of Science and Technology, Griffith University, Brisbane, Australia, 4111.)

Dr Michael McCabe is reader in biochemistry at the James Cook University of North Queensland. Educated in London, he has previously held teaching and research positions in Hong Kong and Singapore. His research interests range from the development of electrodes for medical and veterinary applications to the study of the effect of pesticides and herbicides upon immune responses. (Department of Chemistry and Biochemistry, James Cook University of North Queensland, Townsville, Australia, 4811.)

Dr Rahmah Mohamed is senior lecturer in the Department of Biochemistry, Universiti Kebangsaan Malaysia. Her research interests include the isolation and characterisation of the melioidosis toxin and mammalian systems to assay cytotoxicity. (UKM, Bangi, Selangor, Kuala Lumpur, Malaysia, 43600.)

Mr Geoffrey Newman-Martin is principal chemist with the Commonwealth Department of Community Services and Health in Canberra, Australia. He is also the Scientific Director of the (Australian) National Poisons Information Service and a Captain in the Royal Australian Army Medical Corps. He has been responsible for the development of the National Poisons Register for Australia. (Commonwealth Department of Community Services and Health, GPO Box 9848, Canberra, Australia, 2601.)

Dr Graham M. Nicholson is a lecturer in biological and biomedical sciences at the University of Technology, Sydney. A neuropharmacologist, his major interest involves electrophysiological studies on the neurotoxic actions of funnel-web spider and common brown snake venoms. (School of Biological and Biomedical Sciences, University of Technology, Sydney, Australia, 2007.)

Dr Peter B. Oelrichs is a senior scientist at the Department of Primary Industries, Brisbane, Australia. He is a member of the Poison Plants Committee of the Queensland State Government, and has major research interests in the organic toxins of native and introduced flora in Australia, and their effects on stock. (The National Research Centre for Environmental Toxicology, Kessels Road, Coopers Plains, Australia, 4108.)

Professor Baldomera M. Olivera is a world expert on the biochemistry of cone shell toxins. He is from the Department of Biology, University of Utah, Salt Lake City, Utah 84112, USA.

Professor John Pearn is a paediatrician, professor and head of the Department of Child Health within the University of Queensland and Deputy Dean (Clinical) of the Faculty of Medicine. He has research, patient-care, and teaching interests in clinical toxicology, and is a member of the Board of the American College of Tropical Medicine. (Department of Child Health, Royal Children's Hospital, Brisbane, Australia, 4029.)

Dr Geoffrey A. Pietersz is a senior research officer at the Research Centre for Cancer and Transplantation at the University of Melbourne. His interests include the search for new cancer treatments in the field of targeted cytotoxicity. (Department of Pathology, University of Melbourne, Parkville, Melbourne, Australia, 3052.)

Professor Alan A. Seawright is professor of veterinary toxicology in the Department of Veterinary Pathology at the University of

Queensland and is a recognised world authority in veterinary toxicology. He has special interests in the poisoning of animals by toxic plants and the mechanisms of action of the toxic principles they contain. (School of Veterinary Science, University of Queensland, Brisbane, Australia, 4072.)

Mr Michael J. Thurn is a doctoral student at the University of Technology, Sydney. His research themes focus on the biochemistry and immunology of the Australian paralysis tick, *Ixodes holocyclus*. (Department of Cellular Pathology, University of Technology, Westbourne, Gore Hill, Sydney, Australia, 2065.)

Dr Gary T. Vaughan is a research scientist in the CSIRO Centre for Advanced Analytical Chemistry, at the Division of Fuel Technology, Lucas Heights Research Laboratories, Menai, New South Wales. He has research interests in chemical carcinogenesis and occupational hygiene monitoring. (Private Mail Bag 7, Menai, Australia, 2234.)

Dr Dianne J. Watters is a biochemist and senior research officer at the Queensland Institute for Medical Research in Brisbane, Australia. Her major research contributions have been in the field of cell growth and regulation, and on the effects of toxins on cells. She has published extensively in the area of cell surface receptors, and is currently using marine toxins as a research tool to investigate the differentiation of living cells. (Queensland Cancer Fund Research Unit, Queensland Institute of Medical Research, Herston Rd, Brisbane, Australia, 4029.)

Dr Alan N. Whitaker is an associate professor in the Department of Medicine, University of Queensland, at the Princess Alexandra Hospital. His current research interests include the investigation of the haemostatic components of snake venoms. (Princess Alexandra Hospital, Brisbane, Australia, 4102.)

1 DINOFLAGELLATE TOXINS IN AUSTRALIAN SHELLFISH

GRAHAME J. KELLY AND
GUSTAAF M. HALLEGRAEFF

INTRODUCTION

Almost all marine plants are algae. The larger algae ("seaweeds") are conspicuous because of their size, but in fact the greater proportion of plant growth in the ocean is performed by algae too small to be seen with the naked eye. These are the marine microalgae. One division of marine microalgae that is somewhat exotic in that it is only distantly related to most other plants is the dinoflagellate group. A feature of this group is that a number of species produce substances that act as neurological toxins in humans. Since dinoflagellates, together with other marine microalgae, form the base of the marine food chains, then these toxins can be inadvertently consumed by people when eating shellfish that had, in turn, fed on toxic dinoflagellates (Figure 1.1).

Two groups of poisonings by these toxins have been identified: (a) diarrhetic shellfish poisoning (DSP) that causes gastro-intestinal problems, but no serious effects (except there is recent evidence that the toxins promote tumors), and (b) the more potent paralytic shellfish poisoning (PSP) which, in extreme cases, causes death due to respiratory paralysis (Table 1.1).

This problem of shellfish toxicity due to dinoflagellates has been recognized in parts of the Northern Hemisphere since the late 1700s, but Australia was believed to be free of the problem until 1986 when a potentially toxic dinoflagellate, *Gymnodinium catenatum*, was found to be growing in large numbers in estuarine waters that also contained commercial shellfish farms near Hobart, Tasmania (Hallegraeff and Sumner, 1986). Sub-

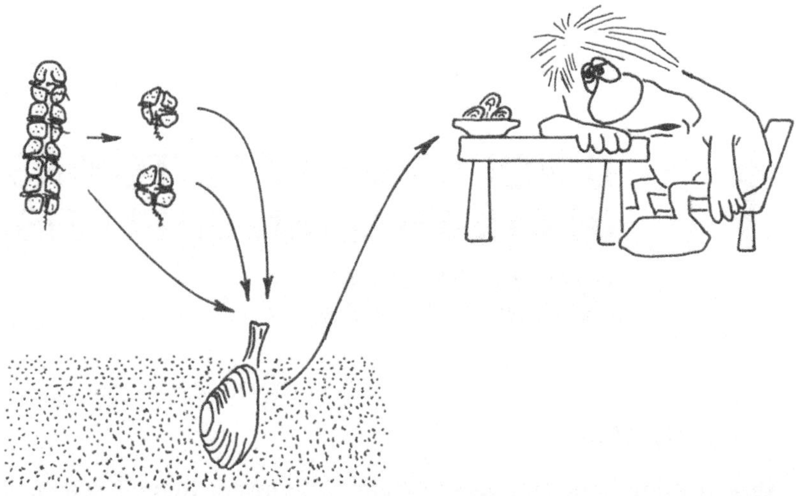

Figure 1.1 Shellfish that feed on dinoflagellates may accumulate toxins that are then passed on to the human consumer.

Table 1.1 Types of shellfish poisoning identified in Australia

	Paralytic shellfish poisoning (PSP)	*Diarrhetic shellfish poisoning (DSP)*
Causative organism(s)	*Gymnodinium catenatum; Alexandrium catenella; Alexandrium minutum*	*Dinophysis acuminata; Dinophysis fortii*
Causative toxin(s)	Saxitoxin and its derivatives	Polyether compounds such as okadaic acid
Symptoms	Tingling and prickly sensations around face, fingers and toes; nausea; in extreme cases muscular paralysis and respiratory difficulty that can lead to death through respiratory paralysis	Abdominal pain, nausea, diarrhoea
Treatment	Pump stomach; artificial respiration	Recovery occurs irrespective of medical treatment

sequently, other toxic species have been identified. A brief description of these microalgae, their toxins, and the control measures required to prevent human poisoning is presented in this report.

THE DINOFLAGELLATES

Gymnodinium catenatum (Figure 1.2a) was the first dinoflagellate able to cause paralytic shellfish poisoning (PSP) to be discovered in Australian waters. It had been linked previously to cases of PSP in Spain and Mexico, and had been recorded in Argentina. Its occurrence in the vicinity of southern Tasmanian oyster and mussel farms raised the question of why cases of PSP apparently had not been recorded in Tasmania. This question became more pressing after analysis of the shellfish revealed that they contained extremely high levels of potential PSP toxins, and action was taken to temporarily close 15 shellfish farms, much to the disdain of shellfish farmers who consumed their product without any apparent ill effects. A resolution of these conflicting observations was obtained from a detailed analysis of the *G. catenatum* toxins using high performance liquid chromatography (Oshima *et al.*, 1987; see below). This analysis revealed that the PSP toxins were of low potency, but could be converted to related, highly toxic PSP toxins during preparation and cooking of seafoods.

Following the discovery of *G. catenatum* in 1986, two further dinoflagellates that contained PSP toxins were found in Port Phillip Bay, Melbourne (*Alexandrium catenella*) and Port River, near Adelaide (*Alexandrium minutum*) (Hallegraeff *et al.*, 1988a). When wild mussels from these areas were examined they were found to contain PSP toxins.

A second group of dinoflagellate toxins that can be passed on to humans through consumption of contaminated shellfish is the polyether compounds that cause diarrhetic shellfish poisoning (DSP). These toxins have been found in the dinoflagellates *Dinophysis acuminata* and *Dinophysis fortii* (Figure 1.2b), which periodically form dense blooms in southern Tasmanian and New Zealand waters, and may have caused DSP since the toxins have been detected in Tasmanian mussels (Hallegraeff, 1987). However, DSP cases are difficult to confirm because they are easily confused with bacterial gastric infections. Their presence is now viewed with more concern because there is now good evidence

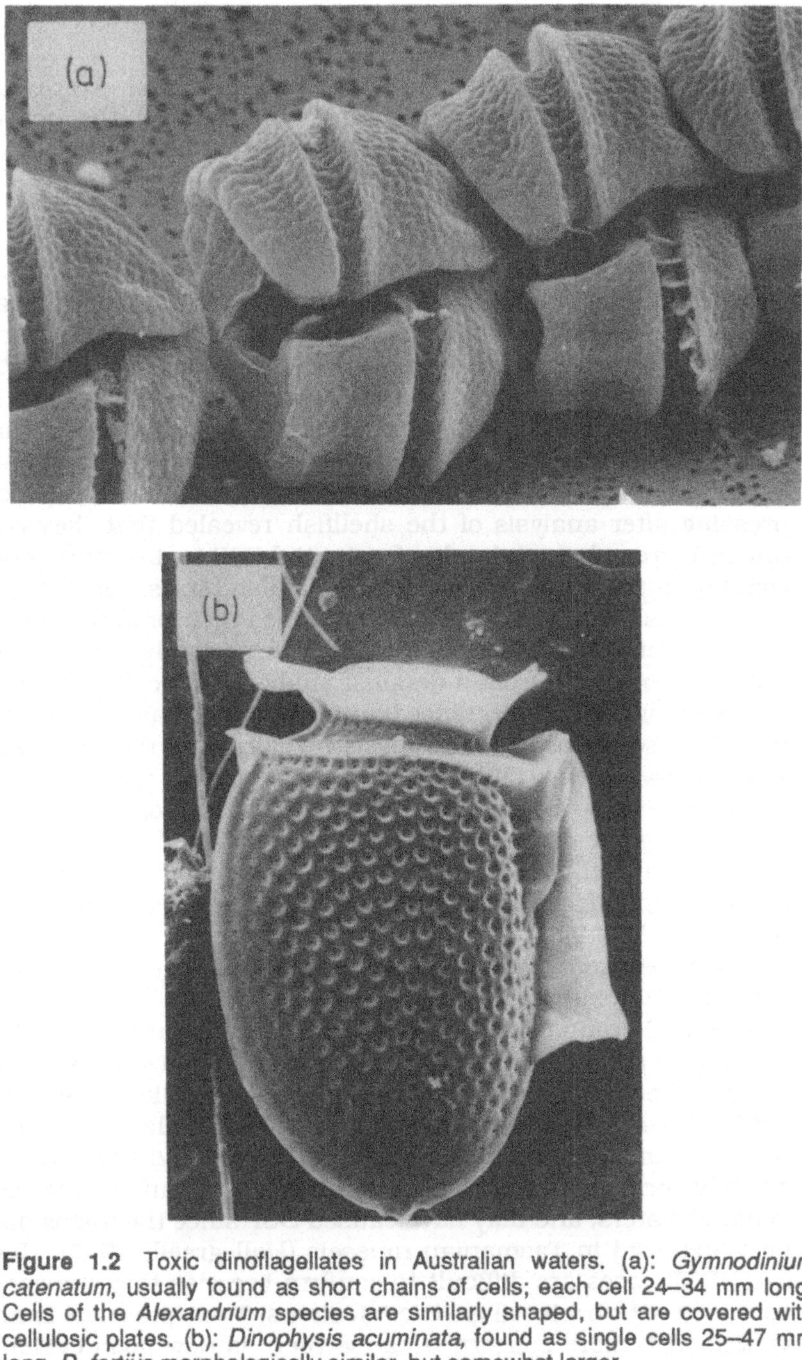

Figure 1.2 Toxic dinoflagellates in Australian waters. (a): *Gymnodinium catenatum*, usually found as short chains of cells; each cell 24–34 mm long. Cells of the *Alexandrium* species are similarly shaped, but are covered with cellulosic plates. (b): *Dinophysis acuminata*, found as single cells 25–47 mm long. *D. fortii* is morphologically similar, but somewhat larger.

that, besides causing gastro-intestinal problems, they also are capable of promoting tumors (Suganuma *et al.*, 1989).

THE TOXINS

PSP toxins are relatively small molecules having some resemblance to guanine; all PSP toxins possess a guanidino group. The predominant PSP toxin in natural populations of *Gymnodinium catenatum* is sulfocarbamoyl gonyautoxin IV (Figure 1.3), so named because (a) the R_4 carbamoyl ($-CONH_2$) group is extended

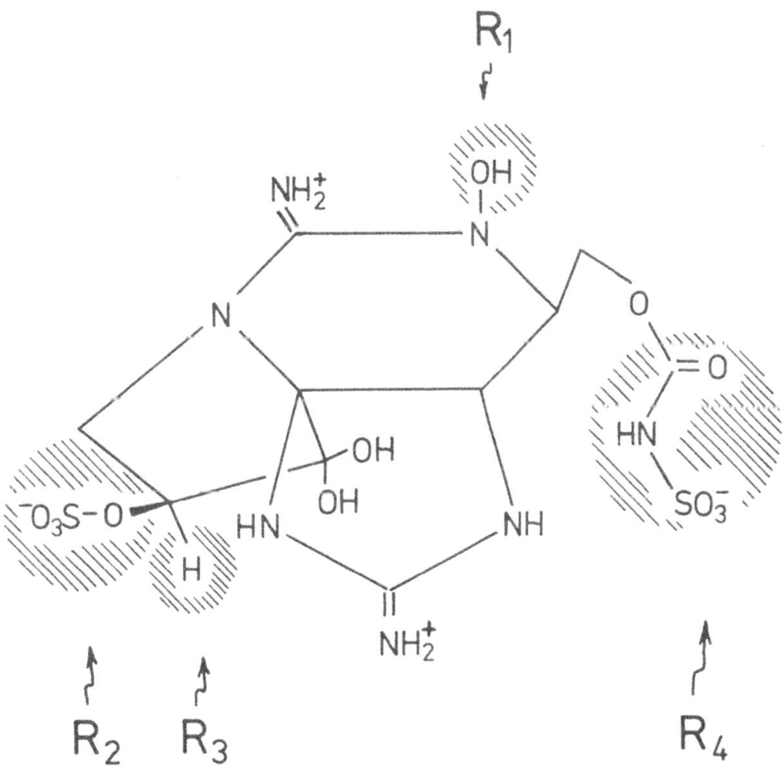

Figure 1.3 Sulfocarbamoyl gonyautoxin IV, currently the major PSP toxin known to be present in Australian waters. The four shaded R-groups are those parts of the molecule that differ between the 18 known derivatives of saxitoxin (see test): R_1 is -H or -OH, R_2 and R_3 are -H or $-OSO_3$, and R_4 is -H, $-CONH_2$ or (as shown) $-CONHSO_3$. For saxitoxin, R_1, R_2 and R_3 are -H, and R_4 is $-CONH_2$. The naming and structures of toxins are given in Oshima *et al.* (1987) and Hallegraeff (1989).

by the addition of $-SO_3$, and (b) the parent compound, gonyautoxin IV, was first isolated from a dinoflagellate of the genus then named *Gonyaulax* (but now renamed *Alexandrium*).

Altogether, some 18 toxins have been identified worldwide. All are considered to be derivatives of saxitoxin, the first PSP toxin characterized (see legend to Figure 1.3)—saxitoxin is named after the butter clam *Saxidomus gigantea*, from which it was isolated for characterization in 1957.

Besides sulfocarbamoyl gonyautoxin IV, *Gymnodinium catenatum* contains lesser amounts of sulfocarbamoyl gonyautoxin I (structure as in Figure 1.3, except the groups on R_2 and R_3 are reversed), and epigonyautoxin VIII (also with this reversal, but with -H at R_1). The toxins in *Alexandrium catenella* are somewhat similar, but *Alexandrium minutum* contains predominantly gonyautoxins I and IV, i.e. with structures as above for the sulfocarbamoyl gonyautoxins I and IV, but lacking the $-SO_3$ in the R_4 group (see Figure 1.3).

It is not known why PSP toxins appear to have no effect on the shellfish that consume the dinoflagellates. The toxins accumulate in the hepatopancreas and gonads of the shellfish, and are then passed on to human consumers. Human poisoning results when toxin molecules block Na^+ channels: the positively charged guanidino group interacts with a negatively charged carboxyl group at the mouth of the Na^+ channel, on the extracellular side of plasma membranes of nerve and muscle cells (Figure 1.4). This mode of action is similar to that of tetrodotoxin, a related toxin produced by certain animals including the puffer fish and the blue-ringed octopus.

As may be expected, the variously structured dinoflagellate PSP toxins interact differently with the Na^+ channel, and therefore display different toxicities. Relatively speaking, all are very toxic: as little as 2 mg of any of the toxins lacking the $-SO_3$ in the R_4 group can be fatal to an adult person. However, the so-called sulfocarbamoyl toxins (with the $-SO_3$ in the R_4 group, as in Figure 1.3) are less toxic, possibly because the negative charge at this position interferes with interaction between the toxin's positively charged guanido group and the Na^+ channel's negatively charged carboxyl group (see Figure 1.4). Nevertheless, these toxins are potentially dangerous because the $-SO_3$ can be hydrolyzed from this R_4 group, thereby converting the toxin to its more potent carbamate partner. The resultant increase in toxicity is 5- to 12-fold (with -H as the R_3 group) or 50- to 90- fold (with $-OSO_3$

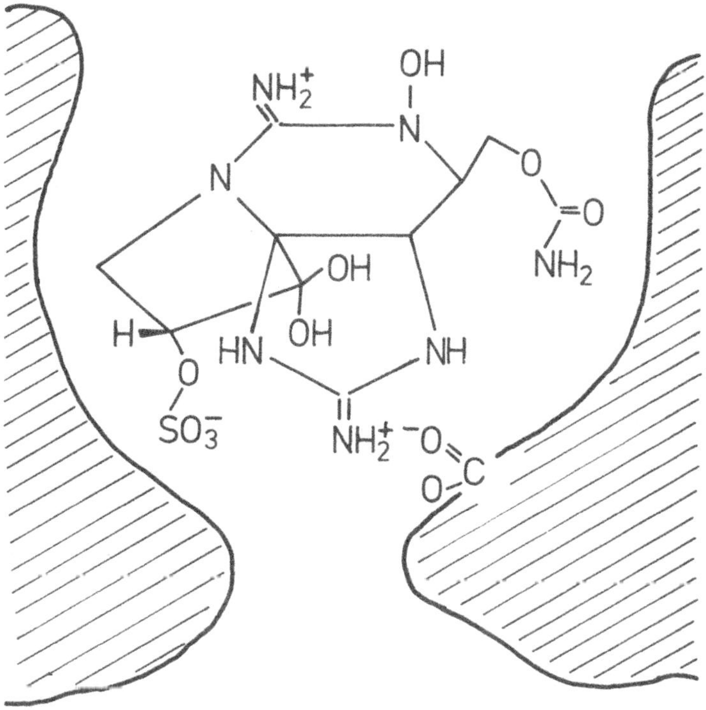

Figure 1.4 A diagrammatic representation of the interaction between a dinoflagellate PSP toxin (in this case gonyautoxin I) and a Na^+ channel. The position of the Na^+ channel's carboxyl group is hypothetical.

as the R_3 group) (Oshima *et al.*, 1988). As pointed out above, Australian PSP toxins are predominantly of the sulfocarbamoyl type, thus it is the potential conversion to more potent toxins that must be accommodated when designing regulations in the interests of public health.

Unlike PSP toxins, DSP toxins have been less well characterized. Their structures, elucidated only during the past few years, place them in the group of fat-soluble polyether compounds. This group also included ciguatera toxins that originate in the dinoflagellate *Gambierdiscus toxicus* (see chapter by M. Capra and J. Cameron in this volume).

The major DSP toxins in Tasmanian *Dinophysis acuminata* and *Dinophysis fortii* have not been investigated, but they are probably okadaic acid and dinophysistoxin-1 (Figure 1.5) since these

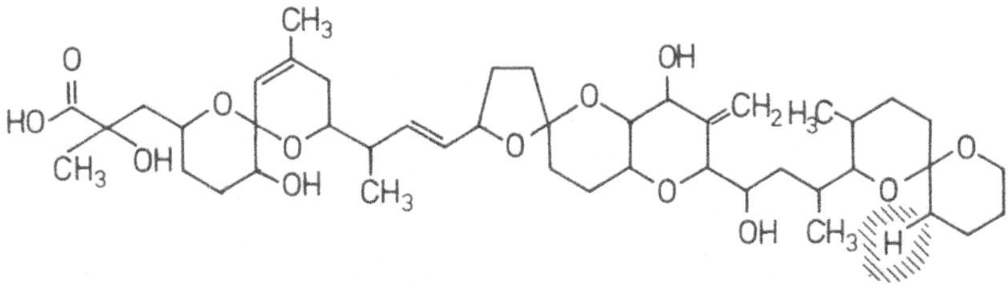

Figure 1.5 The DSP toxin okadaic acid. Dinophysistoxin-1 is identical, except that the shaded hydrogen is replaced by a methyl group. Okadaic acid is named after the sponge *Halichondria okadai* from which it was first isolated, although the source of the toxin was most probably the associated dinoflagellate *Prorocentrum lima*.

are the main toxins in contaminated Tasmanian mussels (Hallegraeff, 1989). These toxins are powerful inhibitors of certain protein phosphatases (Takai *et al.*, 1987), and it is considered likely that this is their primary mode of action, at least with respect to tumor promotion (Suganuma *et al.*, 1989).

CONTROL MEASURES

There is no known practical method for influencing the abundance of toxic dinoflagellates in coastal waters. Shellfish toxicity is therefore managed by (1) attempting to predict the occurrence of dinoflagellate "blooms", (2) continually monitoring the toxin content of shellfish from commercial farms, and temporarily closing any farms when PSP toxin contents exceed 80 mg saxitoxin-equivalent per 100 g shellfish meat, and (3) preventing the spread of toxic dinoflagellates to unaffected areas. This latter measure is particularly important because dinoflagellates form seed-like cysts which allow them to survive for long periods in conditions totally unsuitable for growth. It is not inconceivable that dinoflagellates containing PSP toxins were brought to Australian waters as cysts in the ballast water of cargo vessels (Hallegraeff *et al.*, 1988b).

REFERENCES

Hallegraeff, G. (1987) Red tides in the Australasian region. *CSIRO Marine Lab. Report* **187**, 14 pp.

Hallegraeff, G. (1989) Dinoflagellate toxins in Australian shellfish. In: *Foodborne Microorganisms of Public Health Significance*, edited by K.A. Buckle. p. 366. Pymble: Aust. Inst. Food Sci. Tech.

Hallegraeff, G. and Sumner, C. (1986) Toxic plankton blooms affect shellfish farms. *Aust. Fish.* **45**, 15.
Hallegraeff, G., Steffensen, D.A. and Wetherbee, R. (1988a) Three estuarine Australian dinoflagellates that can produce paralytic shellfish toxins. *J. Plankton Res.* **10**, 533.
Hallegraeff, G., Bolch, C., Koerbin, B. and Bryan, J. (1988b) Ballast water a danger to aquaculture. *Aust. Fish.* **47**, 32.
Oshima, Y., Hasegawa, M., Yasumoto, T., Hallegraeff, G. and Blackburn, S. (1987) Dinoflagellate *Gymnodinium catenatum* as the source of paralytic shellfish toxins in Tasmanian shellfish. *Toxicon* **25**, 1105.
Oshima, Y., Sugino, K. and Yasumoto, T. (1988) Latest advances in HPLC analysis of paralytic shellfish toxins. In: *Proc. 7th IUPAC symposium on Mycotoxins and Phycotoxins*, edited by Natori, S., Hashimoto, K. and Ueno, Y., p. 319. Amsterdam: Elsevier.
Suganuma, S., Suttajit, M., Suguri, H., Ojika, M., Yamada, K. and Fujiki, H. (1989) Specific binding of okadaic acid, a new tumor promoter in mouse skin. *FEBS Lett.* **250**, 615.
Takai, A., Bialojan, C., Troschka, M. and Rüegg, J.C. (1987) Smooth muscle myosin phosphatase inhibition and force enhancement by black sponge toxin. *FEBS. Lett.* **217**, 81.

2 CONESHELL VENOMS—SYNTHESIS AND PACKAGING

DAVID MAGUIRE AND JOYCE KWAN

INTRODUCTION

Peptide toxins are constituents of some of the most potent venoms in the biological world. The lowest LD_{50} values reported for selected toxins from snakes (Chang and Lee, 1963; Eterovic *et al.*, 1975) and spiders (Vicari *et al.*, 1965) are in the range of a few nanomoles of purified toxin per kilogram body weight of test animal. The numerous peptide toxins which have been isolated from snakes, scorpions and spiders have been invaluable in the study of neuromuscular physiology.

In recent years, members of the genus *Conus* have proved a rich ground for the discovery of new physiologically active peptides. In early studies in this area, Endean and coworkers described the pharmacological activities existing in crude extracts from the venom ducts of this large group of marine snails (Whyte and Endean, 1962; Endean and Rudkin, 1963; Endean and Izatt, 1965; Endean, Gyr and Parish, 1974; Endean, Gyr and Surridge, 1979). The range and complexity of activities revealed during those early investigations determined the need for more detailed studies on purified fractions isolated from crude venom extracts. This has been addressed by Olivera's group, and it has been shown that peptides as potent as many of the snake and arachnid toxins can be purified from some *Conus* species (see following article).

The steps involved in the characterization of peptide toxins in these or any other venomous organisms include: purification of the individual toxins which constitute the venom; precise chemical investigations into the structure of isolated toxins; complete chemical synthesis of the structures revealed by those investigations; detailing the precise site of action of those purified toxins and comparing their action with traditional, well described physiological actions; and describing the sites and modes of action of bizarre or previously undescribed actions.

Since there have been relatively few well-documented cases of envenomations by species of *Conus* there are still questions as to which species are toxic to humans. It is important to understand the precise mechanism of venom production since it is possible that some of the final processing of toxin precursors may occur within the prey. The mechanism of venom production can be broadly divided into four separate steps, namely precursor synthesis, packaging, and processing, and venom delivery.

VENOM DELIVERY

All of the extant species of the genus *Conus* use a unique method of delivery of toxin, namely radular teeth which have been modified to resemble harpoons. The microscopic appearance of the tip of one such harpoon is shown in Figure 2.1a as an example of the general appearance of the device. There are interspecies differences in terms of shape and size of these devices, as detailed by Marsh (1977) and James (1980). Toxin is delivered from a venom duct via the hollow shaft of the harpoon formed by curling of the radular teeth edges. The pressure created by contraction of a muscular bulb at the other end of the duct assists in the ejection of venom.

A common feature of the crude venom extracted from all *Conus* species is the presence of numerous granules which are big enough to be observed by low power light microscopy. In some species, granules differ in light-microscopic appearance in different regions of the venom duct. In order to gain some insight into the mechanisms of toxin precursor production and subsequent packaging/processing steps, we have focussed our attention upon these granules.

ELECTRON MICROSCOPY

When venom ducts are prepared for and examined by scanning electron microscopy, small baked-bean shaped granules are

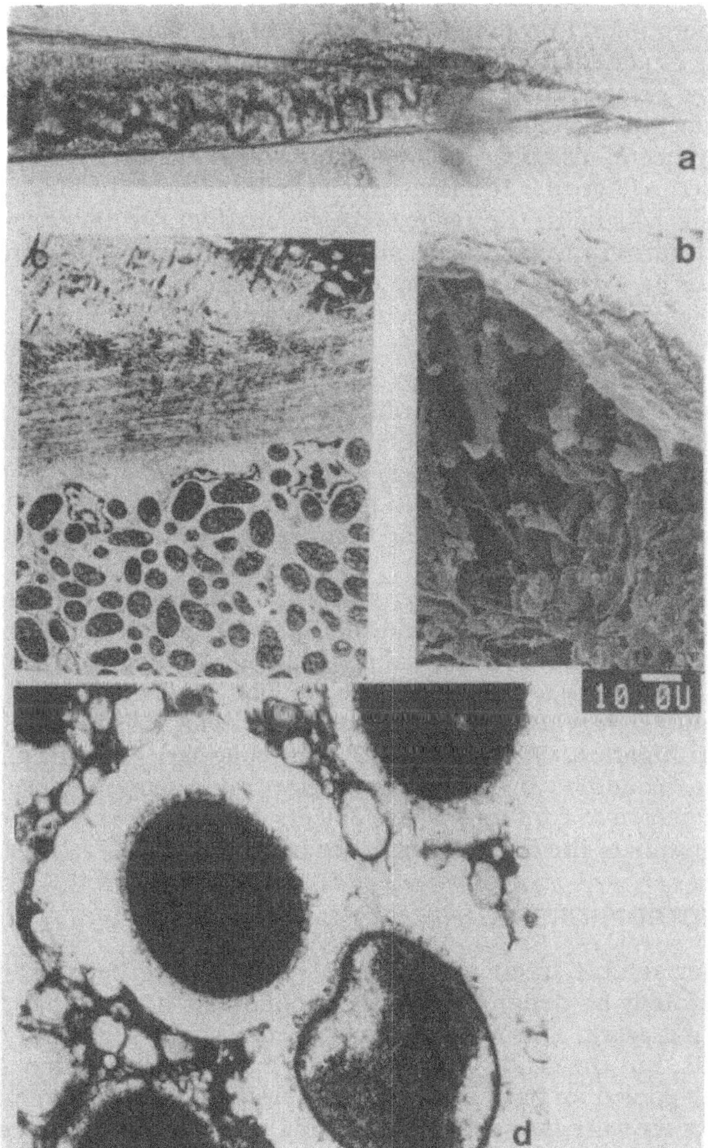

Figure 2.1 Microscopy of some of the components of the *Conus* venom apparatus. (a) Light micrograph of the barb region of a harpoon. (b) Scanning electron micrograph of a section through a duct. (c) Transmission electron micrograph of a section through a duct. (d) Transmission electron micrograph of a section through granules.

numerous and prominent (Figure 2.1b). An examination of portions of venom ducts prepared for transmission electron microscopy reveals that these electron-dense granules are located intracellularly (Figure 2.1c). Another feature of the granules is the existence of an apparent membrane which encloses the granule contents (Figure 2.1d). Because of the fragility of the single layer of cells which line the inside of the venom duct, we have been able to obtain granules essentially free of soluble protein contaminants by extensive washing of crude venom precipitates. Granules prepared in this way can be collected on membrane filters and examined by a number of techniques including scanning electron microscopy (Figure 2.2a). Apart from the features observed in whole duct preparations, some of the purified granules exhibit pitted surfaces (Figure 2.2b). The degree of pitting in any segment preparation depends on its position along the duct with a high proportion of smooth granules in segments close to the venom bulb. This observation is consistent with previous suggestions that there are different regions within the venom ducts.

The purified granules are resistant to disruption by most solvents. However, it was found that granules could be solubilized using strongly chaotropic salts such as guanidine hydrochloride (6M). When this technique was applied to one of the species examined (*Conus textile*), a precipitate was still present after centrifugation. This precipitate was collected on a membrane filter, examined by scanning electron microscopy (Figure 2.2c) and shown to contain sac-like objects. We believe these to be the remnants of the outer membranes of granules.

ELECTROPHORESIS

Water-soluble fractions of *Conus* venoms have been examined previously by denaturing gel electrophoresis (Cruz *et al.*, 1976). In that work, extracts were made from small segments of the ducts, but the crude venom from each segment was then apparently pooled for each animal before electrophoretic examination. Since we were interested in detecting any evidence of processing along the duct, each duct was cut into ten or more segments and the venom extruded from each segment was examined separately (Figure 2.3a). The gel concentration used in this example is suitable for separating low molecular weight proteins. In other gels, not shown here, conditions designed for separating higher molecular weight proteins were used.

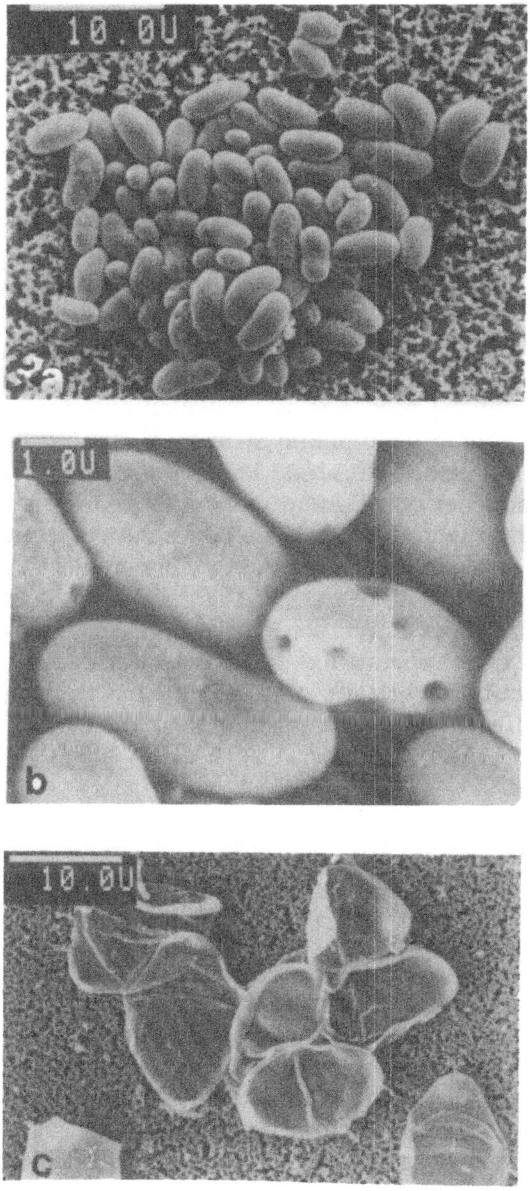

Figure 2.2 Microscopy of isolated granule preparations. (a) Scanning electron micrograph of granules collected on a filter. (b) Higher magnification of this preparation. (c) Scanning electron micrograph of precipitated material collected after disruption of granules.

From all the gels examined, it is apparent that there are three main regions within the duct of each of the species examined. However, it is not yet clear whether these regions reflect different degrees of processing activity, different mechanisms of processing activity or different gene expression. Other possible explanations include a temporally phased arrangement. Thus, one region might be actively producing immature granules while other regions are processing or storing mature granules.

When the guanidine hydrochloride solutions of granules were examined by chaotropic electrophoresis (Figure 2.3b), only a single protein band was observed in extracts from most segments of the ducts of the species examined. This might be interpreted as indicating the presence of a purified precursor protein. However, amino acid sequencing studies of extensively washed granules did not yield a clean sequence. It is possible that membrane proteins or other contaminants were still present in the preparation. Alternatively, a single band on chaotropic gels might mask a mixture of different proteins of similar molecular weights. The guanidine hydrochloride used to disrupt granules can be removed by dialysis. For some duct regions of some species, this leads to the reappearance of particulate matter. The solutions and particulate fractions obtained in this way may prove to be useful in the further examination of *Conus* toxin synthesis.

SUMMARY

The results of the investigations reported here are further evidence that a complex series of processing and packaging events occur in the venom duct of coneshells. It has previously been shown that different pharmacological activities resided in different duct regions (Endean and Izatt, 1965; Endean and Rudkin, 1965). Marsh (1971) showed that caseinase activity was limited to the anterior duct region of the three species studied. The results of our electrophoretic examinations of crude venom from sequential duct regions are also consistent with a regional localization of activities within venom ducts.

A membrane encasing the venom duct granules is revealed by electron microscopic studies. A mechanism must exist to enable the translocation of toxin precursor protein(s) across this barrier. In other organisms, such a process is generally accompanied by the excision of a leader peptide, or signal sequence. Later in this book, David Hillyard presents evidence for such a sequence,

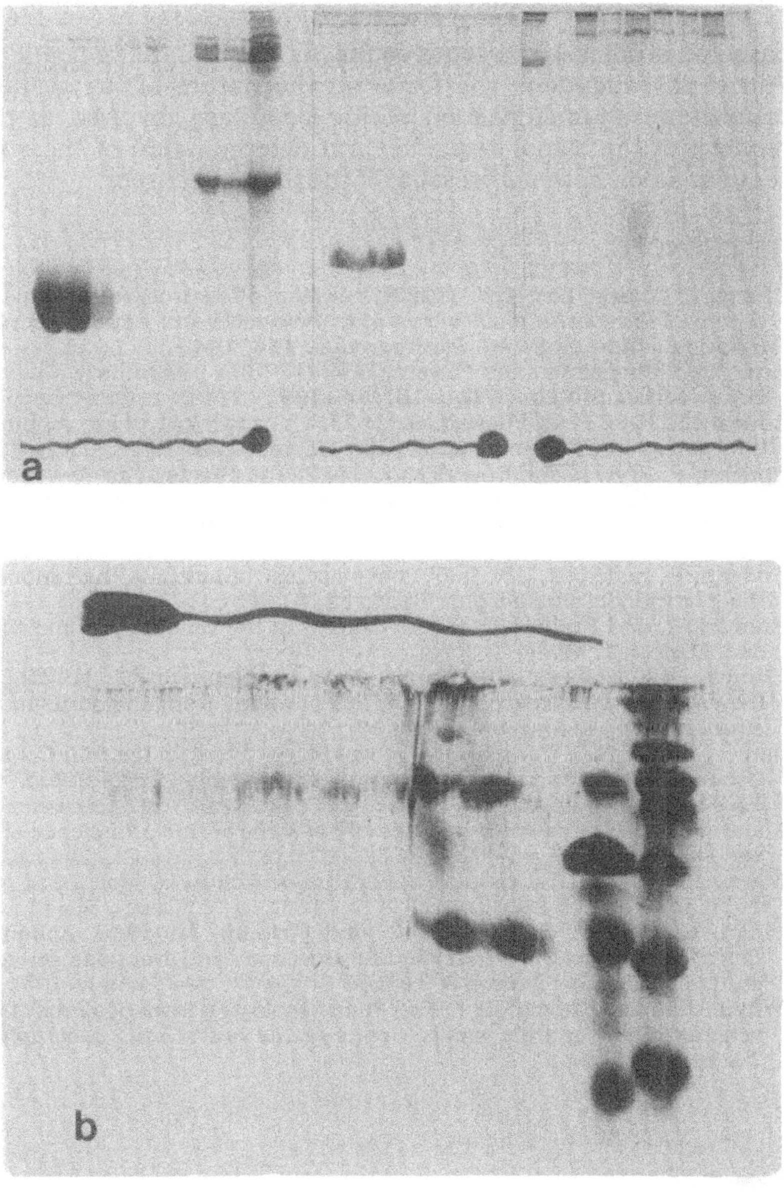

Figure 2.3 Electrophoretogram of *Conus* venom extracts. (a) Soluble fractions from three species of Conus examined by SDS-PAGE (16.5% gels). Individual tracks correspond to adjacent segments of duct shown aligned below each gel. (b) Isolated and disrupted granules collected from adjacent segments of a *Conus* duct. The two lanes on the right are molecular weight markers.

based on studies of the genes coding for the King-Kong peptides. Future investigations will focus on the nature of the granule membrane, characterization of the peptidase involved in the removal of the signal sequence and determination of the subsequent steps in the processing of the toxin precursor.

REFERENCES

Chang, C.C. and Lee, C.Y. (1963) Isolation of neurotoxins from the venom of *Bungarus multicinctus* and their modes of neuromuscular blocking action. *Arch. Int. Pharmacodyn.* **144**, 144.

Cruz, L.J., Corpuz, G. and Olivera, B.M. (1976) A preliminary study of *Conus* venom protein. *Veliger* **18**, 302–308.

Endean, R., Gyr, P. and Parish, G. (1974) Pharmacology of the venom of the gastropod *Conus magus*. *Toxicon* **12**, 117–129.

Endean, R., Gyr, P. and Surridge, J. (1979) The effects of crude venoms of *Conus magus* and *Conus striatus* on the contractile response and electrical activity of guinea-pig cardiac musculature. *Toxicon* **17**, 381–395.

Endean, R. and Izatt, J. (1965) Pharmacological study of the venom of the gastropod *Conus magus*. *Toxicon* **3**, 81–93.

Endean, R. and Rudkin, C. (1963) Studies of the venoms of some conidae. *Toxicon* **1**, 49–64.

Eterovic, V.A., Hebert, M.S., Hanley, M.R. and Bennett, E.L. (1975) The lethality and spectroscopic properties of toxins from *Bungarus multicinctus* (Blyth) toxin. *Toxicon* **13**, 37.

James, M.J. (1980) Comparative morphology of radula teeth in *Conus* : Observations with scanning electron microscopy. *J. Moll. Stud.* **46**, 116–128.

Marsh, H. (1971) The caseinase activity of some vermivorous cone shell venoms. *Toxicon* **9**, 63–67.

Marsh, H. (1977) The radular apparatus of *Conus*. *J. Moll. Stud.* **43**, 1–11.

Vicari, G., Bettini, S., Collotti, C. and Frontali, N. (1965) Action of *Latrodectus mactans tredecimguttatus* venom and fractions on cells cultivated *in vitro*. *Toxicon* **3**, 101.

Whyte, J.M. and Endean, R. (1962) Pharmacological investigations of the venoms of the marine snails *Conus textile* and *Conus geographus*. *Toxicon* **1**, 25.

3 PEPTIDES IN THE VENOM OF THE GEOGRAPHY CONE, *CONUS GEOGRAPHUS*

BALDOMERA M. OLIVERA, DAVID S. JOHNSON, ARYAN AZIMI-ZONOOZ AND LOURDES J. CRUZ

INTRODUCTION

The most venomous gastropod molluscs are the marine snails belonging to the genus *Conus*. A number of human fatalities have resulted from their stings (see Clench and Kondo, 1943; Kohn *et al.*, 1960; Cruz *et al.*, 1985). The approximately 500 species of cones (Walls, 1979) are among the dominant predators in tropical coral reef communities; in tropical marine habitats, the probability of encountering cone snails is fairly high. Despite this, the number of human fatalities is small; stings inflicted on humans are probably defensive. The most deadly species is the geography cone, *Conus geographus*. A high proportion of verified *C. geographus* envenomations are fatal (Yoshiba, 1984). Stinging cases have occurred when the shell of the snail has been unusually perturbed. To pick up a living *Conus geographus* after its shell has been broken (such as by a spear fisherman, for example) appears to be particularly dangerous.

There are reports that other cone snails, including some snail-hunting species such as the Cloth-of-Gold cone, *Conus textile* and the marble cone, *Conus marmoreus*, are also lethal to man (see Cruz, 1990). However, this has not been rigorously confirmed in recent years and since venom collected from the venom ducts of molluscivorous *Conus* is not lethal to vertebrate animals, the veracity of these reports has been questioned (Endean and Rudkin, 1965). However, there is considerable processing of venoms during the injection process. Thus, although the venom collected from the ducts may be inactive on vertebrates, it is conceivable that once processed, snail-hunting *Conus* venoms might be much more potent. This issue needs to be reexamined experimentally, using milked venoms of *Conus textile* and *Conus marmoreus*.

Our laboratory groups have undertaken a systematic examination of the biologically active agents present in *Conus geographus* venom (see Olivera *et al.*, 1985). In this article, we review published work, as well as present some new results that raise intriguing questions about redundancy in the biologically active peptides of *Conus* venoms.

BIOLOGY OF *CONUS GEOGRAPHUS*

Among the ca. 500 species of *Conus*, approximately 10% are fish-hunting, another 10% are mollusc-hunting and the remainder probably hunt worms (mostly polychaetes) (Kohn, 1959; Cruz *et al.*, 1985). All *Conus* species are venomous, and inject their prey using a disposable harpoon-like tooth. The venom is produced in a venom duct, and expelled by a muscular venom bulb (Kohn *et al.*, 1960).

In *Conus geographus*, the mouth (rostrum) of the snail is highly distensible. The geography cone is capable of opening up its mouth so it almost looks like a coral cup (see Figure 3.1). The main strategy for hunting appears to be to stalk sleeping fish; *Conus geographus* is a nocturnal animal, much more active at night than during the day. The snail locates fish by a highly sensitive chemosensory mechanism through its siphon. After locating the fish, it opens up its rostrum as wide as possible very slowly approaching the sleeping fish. If the fish gets close enough, the snail engulfs it with its distensible mouth and stings it to immobilize it quickly. Digestion takes 1–2 hours, after which the snail regurgitates scales, fish bones and the harpoon used to sting the fish. *Conus geographus* is behaviorally a little unusual in terms of its strategy for hunting fish, in that most fish-hunting *Conus* species use a "hook-and-line strategy" while *Conus geographus* basically "nets" its prey before stinging. It appears that in borderline habitats (such as the Marshall Islands), where the species does not grow quite as large as in the center of its range, there seems to be a greater tendency to sting fish before engulf-

Figure 3.1 *Conus geographus* Linne 1758. The top panel shows the shell of *Conus geographus*. Bottom left is a living specimen with its proboscis extended. A snail about to sting defensively would typically have its proboscis exposed as shown. At the bottom right is a *Conus geographus* attempting to feed, with its rostrum distended for catching fish. This species grows to ca. 15 cm.

ing, as do most other fish-hunting *Conus* species (L.J. Cruz, S. Jazwinski and B.M. Olivera, unpublished observations).

Conus geographus is remarkably agile. In aquaria, they have been observed to secrete a strong mucous thread that allows them to levitate themselves downwards in much the same way a spider does with a silk thread. Presumably, they can therefore come off the edges of reefs in order to catch fish and escape from predators. *Conus geographus* has also been observed to eject venom when being held by collectors; this implies that if it is under attack, the snail might eject venom in the direction of the attacking predator even if unable to sting directly.

PARALYTIC COMPONENTS OF *CONUS GEOGRAPHUS* VENOM

At the present time at least 20 different components in *Conus geographus* venom have been purified and characterized. These components may be divided into two general classes, those that directly paralyze fish, and those which are apparently non-paralytic. The vast majority of the biologically active components in *Conus geographus* venom appear to be small peptides, in the range from 5–35 amino acids in length (Cruz *et al.*, 1985; Olivera *et al.*, 1985). However, a number of larger components are also present. For example, a non-paralytic protein with a molecular weight of approximately 13 000 has been described (Clark *et al.*, 1981).

There are three classes of components that are directly paralytic to fish prey ("conotoxins"), and exhibit biological activity on all vertebrate systems tested. These are all disulfide rich peptides in the size range of 13–30 amino acids, typically having a cysteine content of 20–30% (see Cruz *et al.*, 1985; Olivera *et al.*, 1985; Olivera *et al.*, 1988).

The first group of paralytic components characterized were the α–conotoxins (Gray *et al.*, 1981); these peptides inhibit the nicotinic acetylcholine receptor at the neuromuscular junction. Three different molecular forms were described from *Conus geographus* venom, α-conotoxin GI, GII and GIA. The structures of these peptides are shown in Table 3.1. They are 13–15 amino acids in length, and contain two disulfide bonds. It is noteworthy that these toxins have the same receptor target as the major neurotoxins of elapid snakes (such as cobratoxin and α-bungarotoxin). Although α–bungarotoxin is more lethal per molecule at higher doses, α-conotoxins are more rapidly acting than snake toxins, presumably because of their smaller size.

The second group of paralytic toxins are the μ-conotoxins which target to voltage-sensitive Na channels of muscle, and are 22 amino acids in length with three disulfide bonds. In *Conus geographus*, three major forms have been characterized (μ-conotoxins GIIIA, GIIIB and GIIIC) (Sato *et al.*, 1983; Cruz *et al.*, 1985a, 1989). These peptides contain the modified amino acid hydroxyproline. These are the first polypeptide toxins known to compete for binding with the guanidinium toxins, tetrodotoxin and saxitoxin. However, they differ in being much more selective for the skeletal muscle subtype than for axonal Na channels (both inhibited by the guanidinium toxins).

Table 3.1 Structures of some *Conus geographus* peptides. A. Amino acid sequences

Type	*Peptide*	*Sequence*
α-Conotoxin	GI	ECCN<u>P</u>ACGRHYSC*
	GIA	ECCN<u>P</u>ACGRHYSCGK*
	GII	ECCH<u>P</u>ACGKHFSC*
μ-Conotoxin	GIIIA	RDCCT<u>PP</u>KKCKDRQCK<u>P</u>QRCCA*
	GIIIB	RDCCT<u>PP</u>RKCKDRRCK<u>P</u>MKCCA*
ω-Conotoxin	GVIA	CKS<u>P</u>GSSCS<u>P</u>TSYNCCRSCN<u>P</u>YTKRCY*
	GVIIA	CKS<u>P</u>GT<u>P</u>CSRGMRDCCTSCLLYSNKCRRY
Conantokin	G	GEγγLQγNQγLIRγKSN*
Conopressin	G	CFIRNC<u>P</u>KG*
Conotoxin	"GS"	ACSGRGSRC<u>PP</u>QCCMGLRCGRGN<u>P</u>QKCIGAHγDV

Except for γ (γ-carboxyglutamate) and <u>P</u> (trans-4-hydroxyproline), standard one-letter abbreviations for amino acid residues are used. An asterisk indicates that the carboxyl terminus is known to be amidated. The letter designations indicate the species (G, *Conus geographus*) and the particular variant by means of Roman numerals and a capital letter.

The third group of toxins that directly paralyze fish are the ω-conotoxins which inhibit presynaptic voltage sensitive Ca channels (Olivera *et al.*, 1984; Olivera *et al.*, 1985). At least six different molecular forms of ω-conotoxins occur in *Conus geographus* venom, and these can be divided into two groups which are not closely homologous. Although all ω-conotoxins that have been purified from *Conus geographus* venom are paralytic to fish, it is found that the ω-conotoxins belonging to the VI series (ω-conotoxins GVIA, for example), are capable of paralyzing amphibians, while the ω-conotoxins belonging to the VII series (i.e., ω-conotoxins GVIIA, VIIB and VIIC) are not (Olivera *et al.*, 1985). The ω-conotoxins have proven to be useful tools in characterizing neuronal Ca channels, and are even being developed as a diagnostic tool for the Lambert-Eaton Syndrome, an autoimmune disease against Ca channels.

OTHER BIOLOGICALLY ACTIVE PEPTIDES

In addition to the paralytic conotoxins, a number of biologically active peptides which do not by themselves cause paralysis of prey have also been purified. Several of these have been biochemically characterized.

The first non-paralytic peptide described from *Conus geographus* venom was conantokin-G (formerly called conotoxin GV or the "sleeper peptide") (McIntosh *et al.*, 1984; Olivera *et al.*, 1985a; Rivier *et al.*, 1987; Haack *et al.*, 1990). It has recently been shown that this peptide targets to glutamate receptors of the N-methyl-D-aspartate (NMDA) subtype. This is a novel biological activity for a peptide, since no previous examples have been described which exhibit activity against the NMDA receptor. However, the precise specificity of this peptide for NMDA receptors is still being determined. No obvious role is presently known for an NMDA receptor antagonist in fish paralysis, and therefore the biological role of conantokin-G in *Conus geographus* venom remains problematic. An unusual feature of the 17 amino acid long conantokin-G is that the peptide contains five residues of γ-carboxyglutamate ("Gla"), a relatively rare modified amino acid. γ-Carboxyglutamate is post-translationally formed from glutamate by a vitamin K dependent carboxylation system. Before conantokin-G was discovered, no neuroactive peptides were known to contain Gla.

A second non-paralytic peptide from *Conus geographus* venom is conopressin G, a vasopressin analog (Cruz *et al.*, 1987). The structure of conopressin G is shown in Table 3.1; there is little doubt that it is homologous to the vasopressin-oxytocin family of peptides, and its central nervous system effects are at least superficially similar to those induced by vasopressin (see below).

Both conantokin-G and conopressin G were initially detected because of their biological activity in the central nervous system of vertebrates. The screening assay used to detect these activities was i.c.v. injection into mice. Conantokin-G caused sleep-like symptoms in young mice, and a mild hyperactivity in older mice while conopressin G caused immediate scratching upon i.c.v. injection. All neurohypopyseal hormones tested (i.e., vasopressin, oxytocin and other analogs such as vasotocin) also elicit this characteristic scratching syndrome.

ISOLATION OF A NEW γ-CARBOXYGLUTAMATE-CONTAINING PEPTIDE

The successful characterization of conantokin-G and conopressin demonstrated that injection into the central nervous system of mice constitutes a useful screen for biologically active peptides.

In this article we describe two additional peptides detected by such an assay.

We have characterized a second "scratcher" peptide from *Conus geographus* venom. This peptide was discovered while following a convulsant activity which was eluting late (30% acetonitrile in the elution buffer) on a reverse phase semipreparative column well after most of the peptides present in the μ–conotoxin size fraction of the venom had already been eluted (see Cruz *et al.*, 1985a for details). After two more HPLC runs (using analytical columns with progressively shallower gradients), it was found that the convulsant activity was separated from the major absorbance peak, which now exhibited no convulsant activity but instead elicited a pronounced scratching syndrome in mice. Thus, the potent convulsant activity (which appears to be non-peptidic) disguised the scratcher symptoms.

The scratcher peak was subjected to amino acid analysis and a preliminary sequence using an automated sequencer was obtained. The data are consistent with a peptide of 29 amino acids with the following composition: [Ala_2 Arg_1 Asn_1 Asp_1 Cys_4 Gla_1 Glu_1 Gly_3 His_1 Ile_2 Leu_1 Lys_4 Phe_3 Pro_1 Ser_1 Tyr_1 Val_1]. An unusual feature of the scratcher peptide is that it apparently contains one residue of γ–carboxyglutamate (Gla) and two disulfide bonds. Although this peptide causes scratching as does conopressin G, there is no homology detectable between the two.

In addition to the new "scratcher" peptide from *Conus geographus*, another peptide was purified from the μ–conotoxin size fraction of the venom (Cruz *et al.*, 1985a) which also contained γ–carboxyglutamate. The biological activity followed was a marked tendency for mice injected i.c.v. to drag their back legs. After repurification on analytical HPLC, the sequence of this peptide was determined; it was found to be a 34 amino acid peptide with one γ–carboxyglutamate residue. The sequence of this peptide is shown in Table 3.1. A peptide with the same sequence has been previously purified and characterized by Yanagawa *et al.* (1988), using competition with ^{3}H-labeled tetrodotoxin as an assay. These workers have called this peptide conotoxin-CG (a nomenclature adopted in Table 3.1).

Thus, in addition to conantokin-G which contains five residues of γ-carboxyglutamate, there are at least two other peptides in *Conus geographus* venom which have been characterized which also contain this unique amino acid. Both the scratcher peptide, and conantokin-CG have one residue of γ-carboxyglutamate. In contrast to conantokin-G which contains no disulfide bonds, the scratcher peptide has two, and conotoxin-CG has three disulfides. Thus, the disulfide cross-links so characteristic of the major paralytic conotoxins are combined in these two peptides with the γ-carboxyglutamate modification found in conantokin-G.

APPARENT REDUNDANCY OF PEPTIDES IN *CONUS GEOGRAPHUS* VENOM

The peptides which have been sequenced from *Conus geographus* exhibit overlapping physiological specificity at several levels. First, there are three distinct major classes of peptides each capable of paralyzing fish prey. Why have three different classes of paralytic peptides in a single venom? We have argued elsewhere (Olivera *et al.*, 1985) that it may be critically important for the relatively slow-moving cone snails to paralyze their prey as quickly as possible. One reason for having three different paralytic peptides may be to increase the speed of paralysis. In addition, it is more efficient, given a large dilution effect, to have agents that hit three sequential physiological targets, than to try to completely saturate a single target which may be present in great excess in the neuromuscular system.

A second example of overlap are homologs in the venom of the same species. Thus, there are at least three forms of each paralytic conotoxin in *Conus geographus* venom. The significant divergence in sequence of the two classes of ω-conotoxins found in *Conus geographus* venom (as exemplified by ω-conotoxin GVIA and GVIIA) suggests that at least some of the polymorphism may be an attempt to effectively target to a spectrum of related receptor subtypes (either because of phylogenetic or tissue-specific variation).

In the experiments described in this work, a third example of apparent redundancy is seen. We have isolated from *Conus geographus* venom two peptides which both cause mice to scratch, the vasopressin analog conopressin G and the 29 amino acid scratcher peptide described above. We were surprised to find that there is no obvious homology between the two different scratcher activities from the same venom. It is of course possible that different physiological mechanisms underlie scratching activity. Nevertheless, it clearly should not be automatically assumed that peptides causing superficially similar *in vivo* symptomatology will necessarily be homologous to each other.

Finally, it appears that in *Conus geographus* venom, there are two independently evolved peptides which both target to muscle Na channels, and indeed both bind to site 1 of the channel (classically defined by binding of the guanidinium Na channel toxins, tetrodotoxin and saxitoxin). It is remarkable that the μ-conotoxins from *Conus geographus* venom and the conotoxin-CG are, at the present time, the only polypeptide toxins known to target to this particular binding site on the channel. Although apparently unrelated, both peptides have been evolved in this single venom. It is possible that the true biological targets of these toxins differ *in vivo*, i.e., the μ-conotoxins may be specifically targeted to fish prey and conotoxin-CG might have some other function. It is notable that in the last two cases of redundancy, the newly discovered peptides have proven to have γ-carboxyglutamate residues.

Thus, the venom of *Conus geographus* is clearly a complex brew, with functionally overlapping peptides present. Although we can rationalize some of these, the presence of apparently redundant specificities seem an indication that a complex biological milieu provides the driving force for the evolution of the variously targeted small peptides in this remarkable venom.

ACKNOWLEDGEMENT

The work of the authors has been supported by NIH Grant GM22737, and by the Marine Science Institute, University of the Philippines.

REFERENCES

Clark, C., Olivera, B.M. and Cruz, L.J. (1981) A toxin from *Conus geographus* venom which acts on the vertebrate central nervous system. *Toxicon* **19**, 691–699.

Clench, W.J. and Kondo, Y. (1943) The poison cone shell. *Amer. J. Trop. Med.* **23**, 105–120.

Cruz, L.J. (1990) Venomous cones and their toxins. WHO/IPCS monograph on toxins and poisons (in press).

Cruz, L.J., De Santos, V., Zafaralla, G.C., Ramilo, C.A., Zeikus, R., Gray, W.R. and Olivera, B.M. (1987) Invertebrate vasopressin/oxytocin homologs. Characterization of peptides from *Conus geographus* and *Conus striatus* venoms. *J. Biol. Chem.* **262**, 15821–15824.

Cruz, L.J., Gray, W., Yoshakimi, D. and Olivera, B.M. (1985) *Conus* venoms: A rich source of neuroactive peptides. *J. Toxicology-Toxin Reviews* **4**, 107–132.

Cruz, L.J., Gray, W.R., Olivera, B.M., Zeikus, R.D., Kerr, L., Yoshikami, D. and Moczydloweski, E. (1985a) *Conus geographus* toxins that discriminate between neuronal and muscle sodium channels. *J. Biol. Chem.* **260**, 9280–9288.

Cruz, L.J., Kupryszewski, G., LeCheminant, G.W., Gray, W.R., Olivera, B.M. and Rivier, J. (1989) μ–Conotoxin GIIIA, a peptide ligand for muscle Na channels: Chemicals synthesis, radiolabeling and receptor characterization. *Biochemistry* **28**, 3437–3442.

Endean, R. and Rudkin, C. (1965) Further studies of the venoms of Conidae. *Toxicon* **2**, 225–249.

Gray, W.R., Luque, A., Olivera, B.M., Barrett, J. and Cruz, L.J. (1981) Peptide toxins from *Conus geographus* venom. *J. Biol. Chem.* **256**, 4734–4740.

Haack, J.A., Rivier, J., Parks, T.N., Mena, E.E., Cruz, L.J. and Olivera, B.M. (1990) Conantokin-T: A γ–carboxyglutamate containing peptide with NMDA antagonist activity. *J. Biol. Chem.* **265**, 6025-6029.

Kohn, A.J. (1959) The ecology of *Conus* in Hawaii. *Ecol. Monogr.* **29**, 47–90.

Kohn, A.J., Saunders, P.R. and Wiener, S. (1960) Preliminary studies on the venom of the marine snail *Conus*. *Annu. N.Y. Acad. Sci.* **90**, 706–725.

McIntosh, J.M., Olivera, B.M., Cruz, L.J. and Gray, W.R. (1984) γ–Carboxyglutamate in a neuroactive toxin. *J. Biol. Chem.* **259**, 14343–14346.

Olivera, B.M., Gray, W.R. and Cruz, L.J. (1988) Marine snail venoms. In: *Handbook of Natural Toxins*, Vol. **4**, Marine Toxins and Venoms, edited by A. Tu, pp. 327–352, New York and Basel: Marcel Dekker, Inc.,

Olivera, B.M., Gray, W.R., Zeikus, R., McIntosh, J.M., Varga, J., Rivier, J., De Santos, V. and Cruz, L.J. (1985) Peptide neurotoxins from fish-hunting cone snails. *Science* **230**, 1338–1343.

Olivera, B.M., McIntosh, J.M., Cruz, L.J., Luque, F.A., and Gray, W.R. (1984) Purification and sequence of a presynaptic peptide toxin from *Conus geographus* venom. *Biochemistry* **23**, 5087–5090.

Rivier, J., Galyean, R., Simon, L., Cruz, L.J., Olivera, B.M. and Gray, W.R. (1987) Total synthesis and further characterization of the γ–carboxyglutamate-containing "sleeper" peptide from *Conus geographus* venom. *Biochemistry* **26**, 8508–8512.

Sato, S., Nakamura, H., Ohizumi, Y., Kobayahsi, J. and Hirata, Y. (1983) The amino acid sequences of homologous hydroxyproline containing myotoxins from the marine snail *Conus geographus* venom. *FEBS Lett.* **155**, 277–280.

Walls, J.G. (1979) *Cone Shells*, 1011 pp. Hong Kong/Singapore: T.F.H. Publications, Inc. Ltd.

Yanagawa, Y., Abe, T., Satake, M., Odani, S., Suzuki, J. and Ishikawa, K. (1988) A novel sodium channel inhibitor from *Conus geographus*: Purification, structure, and pharmacological properties. *Biochemistry* **27**, 6256–6262.

Yoshiba, S. (1984) An estimation of the most dangerous species of cone shell *Conus geographus* venoms lethal dose in humans. *Jpn. J. Hyg.* **39**, 565.

4 CONESHELL TOXIN GENES

DAVID R. HILLYARD AND SCOTT R. WOODWARD

INTRODUCTION

Cone snails are among the most diverse groups of predators in tropical marine reefs. Their success in this exceedingly competitive habitat stems in large part from their capacity to generate venoms of extraordinary complexity. The biologically active components of these venoms are small peptide molecules (see Cruz *et al.*, 1985; Olivera *et al.*, 1985; Gray *et al.*, 1988; Olivera *et al.*, 1988). Many of these peptides inhibit known receptors of the neuromuscular pathway and are therefore paralytic to prey. Non-paralytic peptide components have also been identified which, curiously, have cross-affinities for receptors in the CNS of higher vertebrate organisms. The identity of these receptor targets and the role of each toxin in *Conus* predation is an area of intense research activity.

The present data base suggests that each of the approximately 500 *Conus* species produces a unique venom cocktail of small peptide molecules (Kohn, 1959). In relatively short geological time, the genus *Conus* has evolved a toxin synthetic capacity of staggering complexity with enormous potential for neurobiological research. Since conotoxins are directly encoded by nucleic acid templates, Cone snails also provide an opportunity to study toxin genomes of great complexity. The initial analysis of contoxin genes from *Conus textile* suggest a powerful role for molecular biology in addressing many questions of conotoxin biology. In addition, the mechanism by which conotoxin diversity has been achieved is itself a question of great interest.

THE CONOTOXIN SYNTHETIC PATHWAY

The biochemical characterization of venom components has focused on the biologically active peptides. From this work it has not been possible to determine whether the toxin molecules are the primary translational products of conotoxin genes or if they arise from precursor "propeptide" molecules. Since many active conotoxins are amidated at their carboxy termini, it has been suspected that at least some conotoxins are derived from larger propeptide molecules (Olivera *et al.*, 1985).

Recently, we isolated cDNA clones for the "King Kong" toxin of *Conus textile* (Hillyard *et al.*, 1988; Woodward *et al.*, 1990). Sequencing of these clones reveals an open reading frame for a 78 amino acid protein. The C-terminal 27 amino acids of this protein correspond to the amino acids of the King-Kong toxin. This King Kong toxin sequence is flanked at its N-terminal border by the basic amino acids lysine and arginine. The N-terminal 20 amino acids of the propeptide are a typical eukaryotic signal peptide sequence required for secretion. The cDNA cloning data therefore suggest that the native King Kong toxin is generated from a precursor protein by proteolytic cleavage at a trypsin-like processing site immediately upstream of the toxin. Recent identification of a number of other cDNA conotoxin clones from *Conus textile* and *Conus geographus* indicates that propeptides are generally intermediates in the conotoxin synthetic pathway. The sequences provided by cDNA cloning will facilitate the direct detection of propeptide molecules in venom extracts.

PROPEPTIDE PRECURSORS AND CONOTOXIN FOLDING

The conformational stability of a typical small conotoxin peptide is achieved by intramolecular disulfide crosslinking between paired cysteine residues. Conotoxins have multiple cysteine residues (most commonly 6) and many different disulfide bond configurations are possible, but only one is biologically active. Indeed, synthetic conotoxins fold *in vitro* to form several differently crosslinked species. For venom peptides to achieve their highest potential activity, cone snails must have a mechanism to form specific disulfide bonds. A logical candidate for this role of folding template is the propeptide molecule itself. Identification of propeptide molecules by cloning will allow direct tests for this folding hypothesis using propeptide molecules expressed from genetically engineered clones.

Analysis of cDNA clones for King Kong propeptides provided additional data to support the folding hypothesis. When DNA probes specific for the N-terminal portion of the King Kong propeptide were used to identify cDNA clones, most of these molecules were found not to encode a propeptide which includes the 27 amino acids of the mature King Kong toxin. Instead, these clones encoded two "King Kong cousin" propeptide molecules KK1 and KK2 as illustrated in Figure 4.1. The N-terminal portion of these propeptide molecules is highly conserved. Also completely conserved is the number and distribution of cysteine residues in the "mature toxin" portion of these putative propeptides. However, none of the other amino acids in the C terminal portion of the molecule has any identity to the known amino acids of the mature King Kong toxin. These clones therefore encode radically different "mature toxin" molecules which are related by a common distribution of cysteine residues and a common excised N-terminal propeptide sequence. We speculate that propeptides of the King Kong class may be able to properly fold a variety of short toxin sequences which share the same cysteine number and distribution as is seen in the King Kong toxin. Recent successes in the identification of other conotoxin cDNA clones will soon test the generality of this phenomenon.

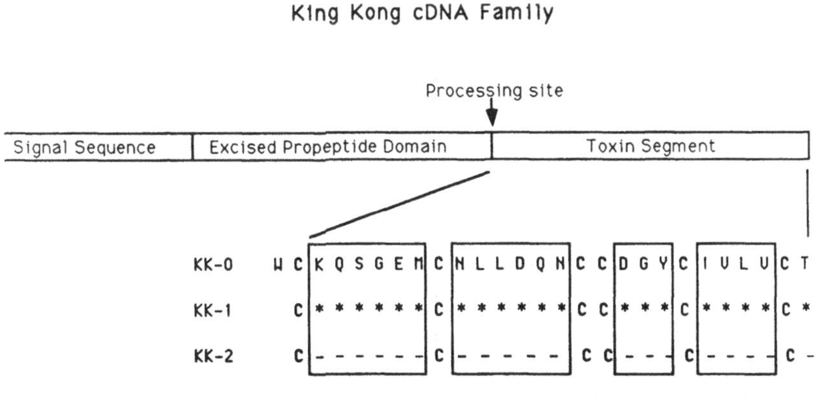

Figure 4.1 The presumptive propeptide structure of proteins encoded by the cDNA clones KK-0, KK-1, KK-2. The putative final processed peptides predicted from potential protease cleavage are shown in the lower portion of the figure.

SUMMARY

The recent identification of cDNA clones for a family of toxins from *Conus textile* has identified the propeptide precursors to mature conotoxin molecules. In addition, a role for propeptide molecules is suggested in the specific folding and directed crosslinking of conotoxins. The generality of these findings for other classes of conotoxin molecules awaits further molecular characterization. In addition to the ideas presented in this paper, it is clear that additional conotoxin cloning will provide answers to a number of specific questions raised by previous peptide sequencing and biochemistry. Included will be the determination of propeptide structures for conotoxins containing "modified" amino acids such as gamma carboxyglutamate. It should be possible to identify amino acid sequences which may signal this (and other) post-translational modification. Further cDNA sequencing should reveal the signals used to direct C-terminal modification of conotoxin peptides, as well as the nature of processing signals for propeptide cleavage to yield the final toxin products. DNA sequence information will be tremendously useful in confirming peptide sequencing results particularly if incomplete or questionable peptide sequence runs are obtained. If other conotoxin cDNA clones are found to fall into families which are unique but share some embedded sequences, then new toxin sequences will be revealed by cDNA sequencing alone. Finally, efforts to identify the structure of conotoxin genomic sequences has the potential for revealing the mechanisms of conotoxin gene diversification.

REFERENCES

Cruz, L.J., Gray, W.R., Yoshikami, D. and Olivera, B.M. (1985) *Conus* venoms: A rich source of neuroactive peptides. *J. Toxicol. Toxin Rev.* **4**, 107–132.

Gray, W.R., Olivera, B.M. and Cruz, L.J. (1988) Peptide toxins from venomous *Conus* snails. *Ann. Rev. Biochem.* **57**, 665–700.

Hillyard, D.R., Olivera, B.M., Woodward, S.R., Corpus, G.P., Gray, W.R., Ramilo, C.A. and Cruz, L.J. (1988) A molluscivorous *Conus* toxin; conserved frameworks in conotoxins. *Biochemistry* **28**, 358–361.

Kohn, A.J. (1959) The ecology of *Conus* in Hawaii. *Ecol. Monogr.* **29**, 47–90.

Olivera, B.M., Gray, W.R. and Cruz, L.J. (1988) Marine snail venoms. In: *Handbook of Natural Toxins*, edited by A.T. Tu, pp. 327-352. New York and Basel: Marcel Dekker, Inc.

Olivera, B.M., Gray, W.R., Zeikus, R., McIntosh, J.M., Varga, J., Rivier, J., De Santos, V. and Cruz, L.J. (1985) Peptide neurotoxins from fish-hunting cone snails. *Science* **230**, 1338–1343.
Woodward, S.R., Cruz, L.J., Olivera, B.M. and Hillyard, D.R. (1990) Constant and hypervariable regions in conotoxin peptides. *Embo. J.* **9**, 1015–1020.

5 TOXINS FROM SEA-SQUIRTS

DIANNE J. WATTERS, CLIFFORD J. HAWKINS AND MARTIN F. LAVIN

INTRODUCTION

Ascidians or sea squirts are marine benthic invertebrate organisms from which a large number of biologically active compounds have been isolated, ranging from simple organic molecules to alkaloids and complex peptides and depsipeptides (Faulkner, 1987). The cyclic depsipeptide compounds, the didemnins, isolated from the colonial ascidian *Trididemnum solidum*, are cytotoxic to tumor cells *in vitro* and have been shown to prolong the life of tumor-bearing animals (Crampton *et al.*, 1984; Jiang *et al.*, 1983) Didemnin B is now in clinical trials in the USA for the treatment of cancer (Stewart *et al.*, 1986). The Great Barrier Reef, with its immense variety of animal life, is a rich source of new cytotoxic compounds with potential antineoplastic activity. Our group has isolated several new interesting cyclic peptide toxins from two species of didemnid ascidian—*Lissoclinum patella* and *Lissoclinum bistratum* (see Figures 5.1 and 5.2)—collected at Heron Island Reef (Degnan *et al.*, 1989 a,b; Hawkins *et al.*, 1990). In this article we review the structures and cytotoxicity of these compounds, and speculate on their mode of action.

STRUCTURAL CHARACTERIZATION

The cyclic peptides found in *L. patella* can be classified into two main types—the patellamides (octapeptides) and the lissoclinamides (heptapeptides containing proline). Patellamides A, B and C, ulithiacyclamide and ascidiacyclamide have been described by

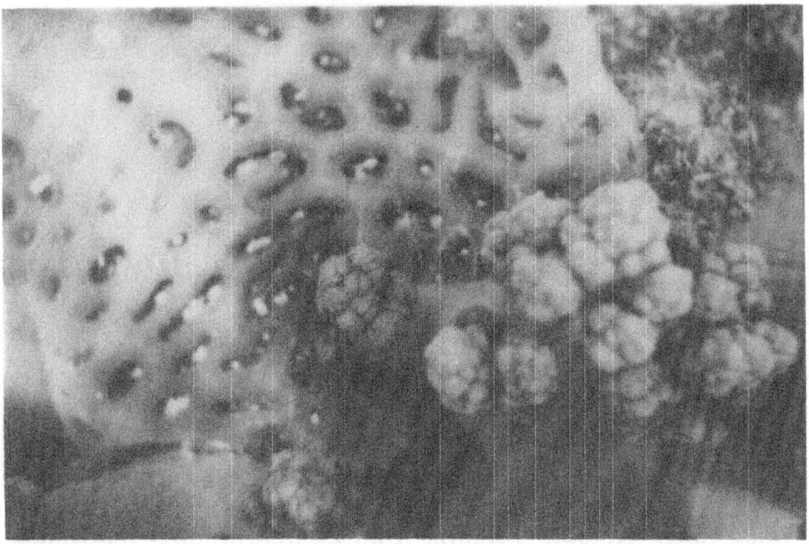

Figure 5.1 A colony of *Lissoclinum patella* growing as a large sheet like structure on the substratum.

Figure 5.2 Several colonies of *Lissoclinum bistratum*.

other researchers (Ireland *et al.*, 1982; Hamamoto *et al.*, 1983). In the lissoclinamide series, ulicyclamide and lissoclinamides 1, 2 and 3 were isolated by Ireland's group. Lissoclinamide 6 has recently been described by Schmitz *et al.* (1989). In *L. patella* from Heron Island Reef we reproducibly find lissoclinamides 4, 5, 7 and 8, which together with 6 establish a nice series for the study of structure-activity relationships. These compounds differ from one another only by minor structural changes e.g. thiazole vs thiazoline rings and/or stereochemistry of the amino acids at positions R_1 and R_2. The thiazole and thiazoline rings arise from the amino acid cysteine and the oxazoline rings from threonine by internal condensation. The left-hand half of the lissoclinamide molecule is invariable with L-threonine, L-proline and L-phenylalanine consistently being found. (See Table 5.1, pages 39–40 for structures of patellamides and lissoclinamides.)

The cyclic peptides are purified from an organic solvent extract of frozen animal tissue by chromatography on a preparative reverse-phase HPLC column (Degnan *et al.*, 1989 a,b). The structures are determined by amino acid analysis, two-dimensional NMR Spectroscopy and mass spectroscopy. Acid hydrolysis followed by derivatization of the amino acids to their N-pentafluoropropionyl isopropyl esters and chiral gas chromatography gives information on the stereochemistry except for amino acids attached to thiazole rings. Difficulties are also encountered with amino acids adjacent to thiazoline rings as these tend to racemise during hydrolysis. For these reasons it is only possible to give a tentative configuration at positions R_1 and R_2 of the lissoclinamides.

In the patellamide series the amino acids adjacent to the thiazole rings are invariably of the D configuration whereas the others are always L. (Determined by X-ray crystallography and comparisons of NMR data.)

L. bistratum extracts contain similar cyclic peptides, however in this species only hexapeptides are found. To date, three have been characterized—bistratamides A and B and cycloxazoline with all amino acids being of the L-configuration. In addition we have isolated from this organism novel macrocyclic polyethers—the bistratenes—which are extremely cytotoxic. (See Table 5.1 pages 39–41 for structures of bistratamides, cycloxazoline and bistratenes.)

CYTOTOXICITY STUDIES

The cytotoxicities of the compounds thus far isolated from *L. patella* and *L. bistratum* are given in Table 5.1. The cytotoxicities were examined *in vitro* using transitional bladder carcinoma cells (T24) and SV40-transformed fibroblasts (MRC5CV1). Cell viability assays were performed using a colorimetric assay which relies on the formation of a blue formazan dye by live mitochondria (Mosmann, 1983). The IC_{50} values represent the concentration of compound required for 50% inhibition of cell growth. The cells were exposed to the toxins for 1 hour, and viability determined after several days. Of the patellamides, ulithiacyclamide is the most potent. It is also the only one containing a disulfide bridge across the ring. Perhaps this bridge holds the molecule in the optimal conformation for biological activity. With the exception of lissoclinamide 7, the lissoclinamides display marginal cytotoxicity. Interestingly, lissoclinamide 7 is exceedingly potent and rivals didemnin B in cytotoxicity. It differs only slightly in structure from lissoclinamides 4, 5, 6 and 8 in having two thiazoline rings instead of one thiazole/one thiazoline or two thiazole rings. The presence of two thiazole rings leads to a considerable decrease in activity (lissoclinamide 5 *vs* lissoclinamide 7). This is interesting in light of the many biologically active compounds

Table 5.1 Cytotoxicity of compounds isolated from *L. patella* and *L. bistratum*. (Structures of patellamides reprinted with permission from *J. Med. Chem.*, **32**, 1349–54. Copyright 1989. American Chemical Society.)

Compound	*IC_{50} (µg/mL)*
Ascidiacyclamide	75
Ulithiacyclamide	0.15
Patellamide D	90
Lissoclinamide 4	10
Lissoclinamide 5	10
Lissoclinamide 7	0.04
Lissoclinamide 8	7
Bistratamide A	60
Bistratamide B	>100
Cycloxazoline	0.5
Bistratene A	0.07
Bistratene B	0.09

Results are reported for T24 bladder carcinoma cells and a 1-hour exposure.

Table 5.1 (Continued)

Lissoclinamides

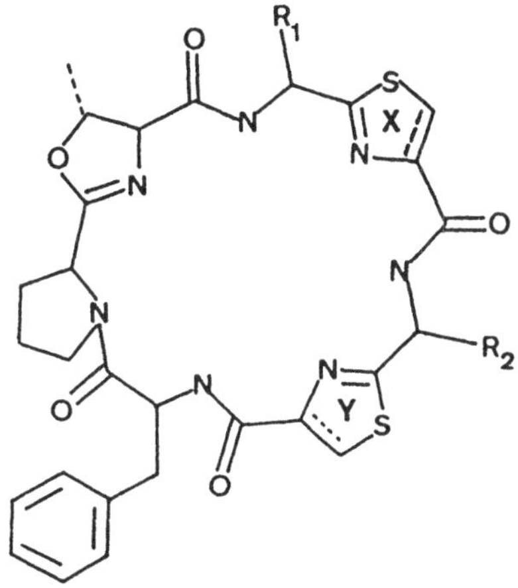

Lissoclinamides	X	Y	R_1	R_2
Ulicyclamide	thiazole	thiazole	L-Ile	D-Ala
Lissoclinamide 1	thiazole	thiazole	L-Val	D-Ile
Lissoclinamide 2	thiazoline	thiazole	D-Ile	D-Ala
Lissoclinamide 3	thiazoline	thiazole	D-Ile	D-Ala
Lissoclinamide 4	thiazoline	thiazole	D-Val	L-Phe
Lissoclinamide 5	thiazole	thiazole	D-Val	L-Phe
Lissoclinamide 6	thiazoline	thiazole	D-Val	D-Phe
Lissoclinamide 7	thiazoline	thiazoline	?-Val	D-Phe
Lissoclinamide 8	thiazoline	thiazole	L-Val	D-Phe

Bistratamides

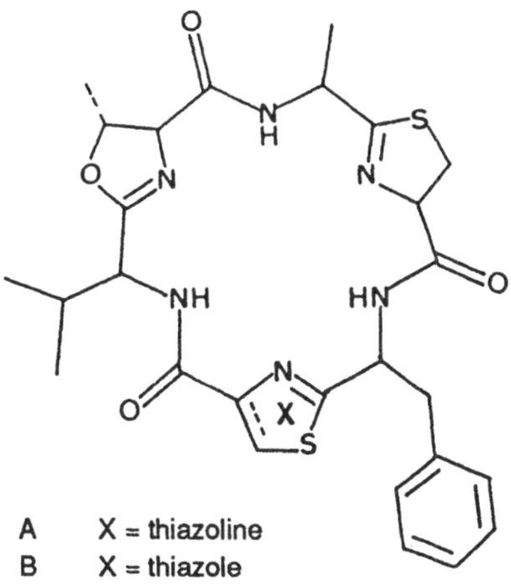

A X = thiazoline
B X = thiazole

Table 5.1 (Continued)

Cycloxazoline

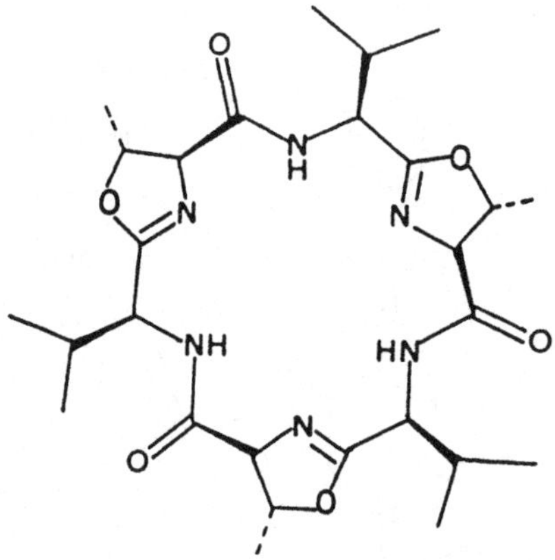

Patellamides

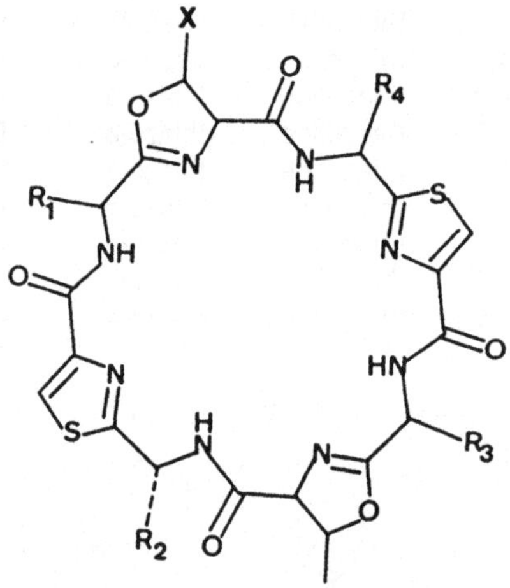

Patellamides	X	R_1	R_2	R_3	R_4
Patellamide A	H	D-Val	L-Ile	D-Val	L-Ile
Patellamide B	Me	D-Ala	L-Leu	D-Phe	L-Ile
Patellamide C	Me	D-Ala	L-Val	D-Phe	L-Ile
Patellamide D	Me	D-Ala	L-Ile	D-Phe	L-Ile
Ascidiacyclamide	Me	D-Val	L-Ile	D-Val	L-Ile
Ulithiacyclamide	Me	D-Leu	L-½Cys	D-Leu	L-½Cys

Table 5.1 (Continued)

Bistratenes

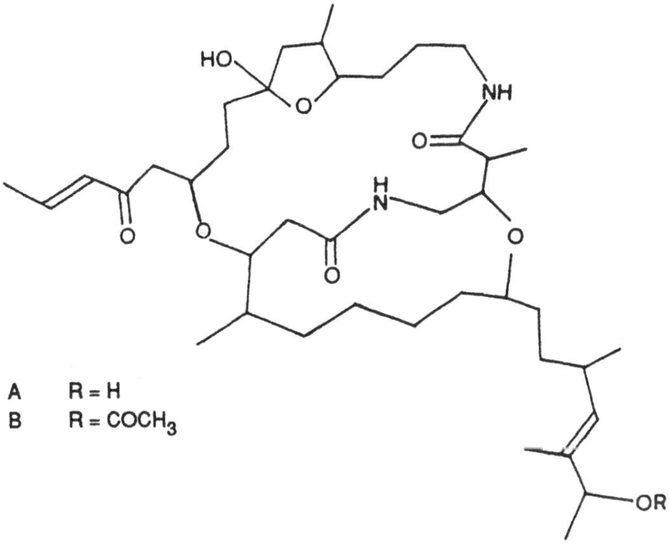

A R = H
B R = $COCH_3$

containing thiazole rings e.g. the dolastatins, (Pettit *et al.*, 1982), thiostrepton (Anderson *et al.*, 1970) and botromycin (Waisvicz *et al.*, 1957). Shioiri *et al.* (1987) have compared the cytotoxicities of a large number of natural and synthetic cyclic peptides and concluded that the oxazoline function is essential for cytotoxicity and that the cyclic skeleton may not be essential. This appears to be substantiated by the fact that cycloxazoline and lissoclinamide 7 are the most potent of the cyclic peptides we have found. On the other hand, dolastatin 10 which does *not* contain a thiazoline or oxazoline ring is the most potent antineoplastic agent presently known. A more likely determinant of cytotoxicity is the overall conformation of the molecule and further speculation awaits the elucidation of the crystal and solution structures of the lissoclinamide compounds.

MODE OF ACTION

The crystal structures of ascidiacyclamide and patellamide D have been determined and also the solution structure of ulithiacyclamide (Hamamoto *et al.*, 1983; Ishida *et al.*, 1989; Schmitz *et al.*, 1989). They are all quite different from each other despite the similarity of their cyclic skeletons. From the structures it is obvious that all the nitrogen atoms point toward the center of the ring. These molecules thus appear to be designed by nature to coordinate metal ions. We have shown that these cyclic peptides are indeed capable of binding metal ions (van den Brenk *et al.*, 1989, 1990). Patellamide D can accommodate 1 or 2 copper or zinc atoms while the smaller hexapeptides, bistratamide A and cycloxazoline can bind only one metal atom. The question arises as to whether this capability of binding metal ions is related to the biological activity of the cyclic peptides. Copper was chosen for study because of the ease of determining metal complex formation by spectroscopic means (circular dichroism and electron spin resonance). However copper complexes are probably not of physiological relevance. There is no difference in cytotoxicity between the copper complexes and the parent compounds so far studied. Further investigations of coordination with other metals are in progress.

Effects on Intracellular Calcium

The effect of these compounds on intracellular calcium levels in mouse thymocytes has been examined. The cells were loaded with a fluorescent calcium indicator Indo-1 and fluorescence was monitored after addition of the test compounds. Preliminary results indicate that lissoclinamides 8 and 7 (7 being more potent) cause a progressive increase in basal intracellular calcium with time (D. Watters, R. Tellam—unpublished observations). The basis for this increase is currently under study. The bistratenes on the other hand have no effect on basal calcium levels in cells. They have some properties in common with the phorbol esters and their mechanism of action appears to be markedly different from that of the cyclic peptides. Bistratenes A and B cause certain changes indicative of the induction of differentiation in a human leukemia cell line, HL-60.

Cell Differentiation

The HL-60 cell line, derived from a patient with acute myeloblastic leukemia, has become an attractive model for studies of human myeloid cell differentiation (Collins, 1987; Birnie, 1988). The cells which resemble promyelocytes can be induced to differentiate terminally *in vitro* into monocyte/macrophage cells by compounds such as the phorbol esters, 1,25-dihydroxyvitamin D3 and sodium butyrate. However the mechanisms by which these agents induce differentiation have remained largely obscure. The phorbol esters e.g. TPA (12-0-tetradecanoylphorbol 13-acetate) are thought to exert their actions by mimicking diacylglycerol in enhancing the calcium and phospholipid-dependent activity of protein kinase C (PKC). However diacylglycerol analogs fail to induce differentiation of HL-60 cells (Morin *et al.*, 1987). Another marine toxin, bryostatin 1, isolated from the bryozoan *Bugula neritina* (Pettit *et al.*, 1982) also enhances the calcium - and phospholipid-dependent PKC activity, but its effects on HL-60 cells are concentration dependent (Stone *et al.*, 1988). Low concentrations (10 nM) induce monocytic differentiation whereas high concentrations (100 nM) inhibit TPA-induced differentiation. Thus there is still considerable controversy over the role of PKC in the differentiation process.

Morphological Studies

When HL-60 cells were treated with 100 nM bistratene A (IC_{50}, 424 nM for these cells), they adhered to the culture plate and acquired a spindle-shaped morphology and prominent pseudopodia (Figure 5.3b). These changes were visible as early as 1 h after addition of the compound. With time (up to 96 h) there was an increase in cell size (Figure 5.3c). TPA-treated cells were more strongly adherent and showed greater clumping (Figure 5.3d).

The presence of monocyte/macrophage differentiation markers on the bistratene treated cells was examined. The treated cells showed some of the characteristics of macrophages, for example adherence to plastic and expression of OKM-1 and Leu-M3 antigens, but did not appear to be activated since they lacked expression of HLA-DR antigens and nonspecific esterase activity (NSE), did not phagocytize latex beads and failed to produce interleukin 1. HL-60 cells treated with TPA on the other hand express NSE and display phagocytosis as do cells treated with

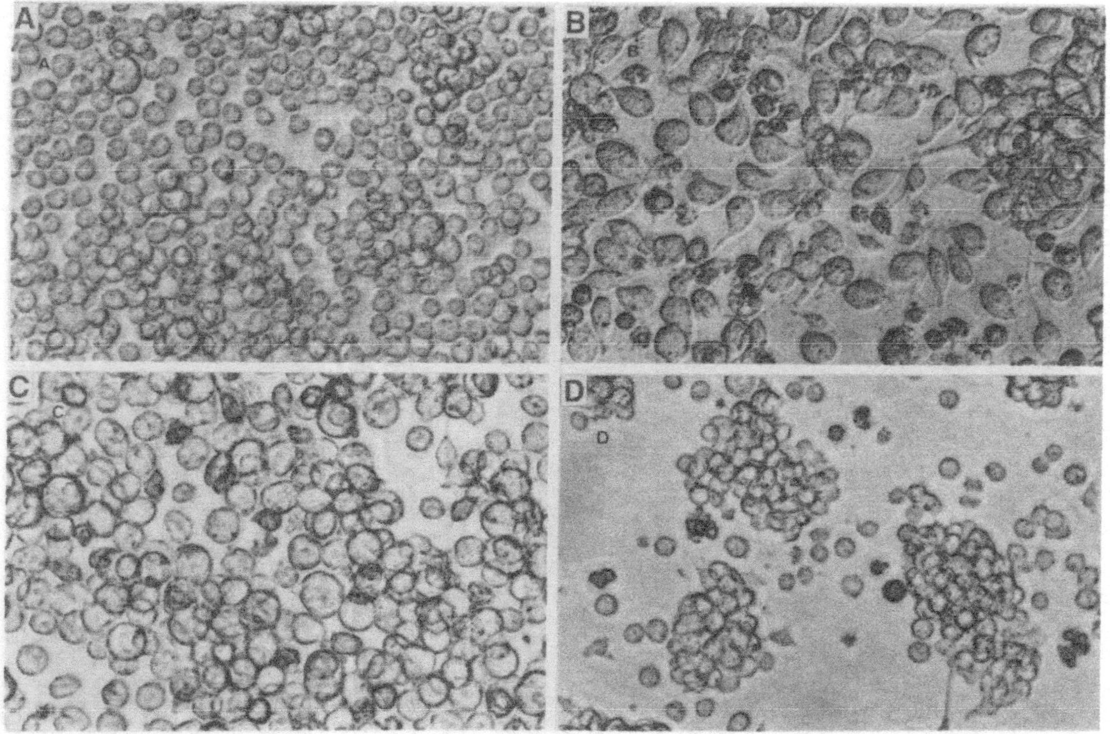

Figure 5.3 (a) HL-60 (human leukemia) cells normally grow in suspension. (b) HL-60 cells treated with 100 nM bistratene A for 24 h. (c) HL-60 cells 96h after treatment with bistratene A. (d) The same cells treated with 16 nM of the phorbol ester TPA for 24 h. (Reprinted with permission from *Biochemical Pharmacology*, **39**, 1609–14. Copyright 1990, Pergamon Press.)

bryostatin 1, indicating a more mature phenotype induced by these agents (Watters *et al.*, 1990).

Protein Kinase C

The effect of bistratene B on the activity of purified type II PKC from bovine spleen was examined in collaboration with Dr Susan Hamilton, Biochemistry Department, University of Queensland. Bistratene B can enhance the phosphatidylserine-dependent activity of PKC although not as dramatically as TPA or the diglyceride, diolein. The concentration required for half-maximal enhancement of activity was 5 μM. This compares with 50 nM required for the half-maximal expression of OKM-1 and Leu M3 antigens (Watters *et al.*, 1990). It is unlikely on this basis that the activation of type II PKC is the primary mechanism by which the bistratenes induce differentiation. It will be of interest to characterize further the interaction of the bistratenes with other PKC isoenzymes since differential effects on the various isoforms have been found with TPA and bryostatin (Kraft *et al.*, 1988; Strulovici *et al.*, 1989).

We have also shown that bistratene A has no effect on DNA synthesis. The primary mode of action (whether a PKC-dependent or PKC-independent process) remains to be determined and this new class of toxins promises to provide us with useful new probes with which to dissect the molecular mechanisms governing cell growth regulation and differentiation.

SYMBIOTIC RELATIONSHIPS

Most ascidians from which biologically active compounds have been isolated are in symbiosis with unicellular algae. For example *T. solidum* has a Cyanophyte symbiont *Synechocystis trididemni* whereas *L. patella* and *L. bistratum* have prokaryotic algal cells of the genus *Prochloron* (Cox, 1986). The origin of the cyclic peptides has not yet been determined with certainty but they are presumed to be synthesized by the *Prochloron*. It is interesting then that *L. patella* and *L. bistratum* with the same symbiont produce different suites of cyclic peptides. The function of the toxic peptides in the ascidian is presumably to avoid predation. Just how the animal escapes the noxious effects itself is intriguing. Substantial quantities (up to 10% dry weight of organic extract) of the very toxic bistratenes are found in *L. bistratum*. The cyclic peptides are present in much lower concentration, typically <1% of dried extract where 100 g frozen animal (wet weight) yields approximately 200 mg of extract.

In conclusion, the two *Lissoclinum* species have yielded a large number of fascinating compounds some of which may eventually find use as potential anti-cancer agents.

REFERENCES

Anderson, B., Hodgkin, D.C. and Viswamitra, M.A. (1970) The structure of thiostrepton. *Nature* **225**, 233.

Birnie, G.D. (1988) The HL-60 cell line: A model system for studying human myeloid cell differentiation. *Br. J. Cancer* **58**, Suppl. IX, 41–45.

Collins, S.J. (1987) The HL-60 promyelocytic leukemia cell line: proliferation, differentiation, and cellular oncogene expression. *Blood* **70**, 1233.

Cox, G. (1986) Comparison of *Prochloron* from different hosts I: Structural and ultrastructural characteristics. *New Phytol.* **104**, 429.

Crampton, S.L., Adams, E.G., Kuentzel, S.L., Li, L.H., Badiner, G. and Bhuyan, B.K. (1984) Biochemical and cellular effects of didemnins A and B. *Cancer Res.* **44**, 1796.

Degnan, B.M., Hawkins, C.J., Lavin, M.F., McCaffrey, E.J., Parry, D.L., van den Brenk, A.L. and Watters, D.J. (1989) New cyclic peptides with cytotoxic activity from the ascidian *Lissoclinum patella*. *J. Med. Chem.* **32**, 1349.

Degnan, B.M., Hawkins, C.J., Lavin, M.F., McCaffrey, E.J., Parry, D.L. and Watters, D.J. (1989) Novel cytotoxic compounds from the ascidian *Lissoclinum bistratum*. *J. Med. Chem.* **32**, 1354.

Faulkner, D.J. (1987) Marine natural products. *Nat. Prod. Reports* **4**, 539.

Hamamoto, Y., Endo, M., Nakagawa, M., Nakanishi, T. and Mizukawa, K. (1983) A new cyclic peptide, ascidiacyclamide isolated from ascidian. *J. Chem. Soc. Chem. Commun.* 323.

Hawkins, C.J., Lavin, M.F., Marshall, K.A., van den Brenk, A.L. and Watters, D.J. (1990) Structure–Activity Relationships in the lissoclinamides: Cytotoxic cyclic peptides from the ascidian *Lissoclinum patella*. *J. Med. Chem.* **33**, 1634.

Ireland, C.M., Durso, A.R., Newman, R.A. and Hacker, M.P. (1982) Antineoplastic cyclic peptides from the marine tunicate *Lissoclinum patella*. *J. Org. Chem.* **47**, 1807.

Ishida, T., Ohishi, H., Inoue, M., Kamigauchi, M., Sugiura, M., Takao, N., Kato, S., Hamada, Y. and Shiori, T. (1989) Conformational properties of ulithiacyclamide, a strongly cytotoxic cyclic peptide from a marine tunicate, determined by ^{1}H-nuclear magnetic resonance and energy minimisation calculations. *J. Org. Chem.* **54**, 5337.

Jiang, T.L., Liu, R.H. and Salmon, S.E. (1983) Antitumor activity of didemnin B in the human tumour stem cell assay. *Cancer Chemother. Pharmacol.* **11**, 1.

Kraft, A.S., Reeves, J.A. and Ashendel, C.L. (1988) Differing modulation of protein kinase C by bryostatin 1 and phorbol esters in JB6 mouse epidermal cells. *J. Biol. Chem.* **263**, 8437.

Morin, M.J., Kreuter, D., Rasmussen, H. and Sartorelli, C. (1987) Disparate effects of promyelocytic leukemia cell differentiation. *J. Biol. Chem.* **262**, 11758.

Mosmann, T. (1983) Rapid colorimetric assay for cellular growth and survival: Application to proliferation and cytotoxicity assays. *J. Immunol. Methods* **65**, 55.

Pettit, G.R., Herald, C.L., Doubek, D.L., Herald, D.L., Arnold, E. and Clardy, J. (1982) Isolation and structure of bryostatin 1. *J. Am. Chem. Soc.* **104**, 6846.

Pettit, G.R., Kamano, Y., Brown, P., Gust, D., Inoue, M. and Herald, C.L. (1982) Structure of the cyclic peptide dolastatin 3 from *Dolabella auricularia. J. Am. Chem. Soc.* **104**, 905.

Schmitz, F.J., Ksebati, M.B., Chang, J.S., Wang, J.L., Hossain, M.B. and van der Helm, D. (1989) Cyclic peptides from the Ascidian *Lissoclinum patella*: Conformational analysis of patellamide D by X-Ray analysis and molecular modeling. *J. Org. Chem.* **54**, 3463.

Shiori, T., Hamada, Y., Kato, S., Shibata, M., Kondo, Y., Nakagawa, H. and Kohda, K. (1987) Cytotoxic activity of cyclic peptides of marine origin and their derivatives: importance of oxazoline functions. *Biochem. Pharmacol.* **36**, 4181.

Stewart, J.A., Tong, W.R., Hartshorn, J.N. and McCormack, J.J. (1986) *Proc. Am. Soc. Clin. Oncol.* **5**, 33.

Stone, R.M., Sariban, E., Pettit, G.R. and Kufe, D.W. (1988) Bryostatin 1 activates protein kinase C and induces monocytic differentiation of HL-60. *Blood* **72**, 208.

Strulovici, B., Daniel-Issakani, S., Oto, E., Nestor, J., Chan, H. and Tsou, A-P. (1989) Activation of distinct protein-kinase C isozymes by phorbol esters: correlation with induction of interleukin 1b gene expression. *Biochem.* **28**, 3569.

van den Brenk, A.L., Hanson, G.R., Hawkins, C.J., Lim, Y.Y. and Watters, D.J. (1989) Metal complexes of cyclic hexapeptides from *Lissoclinum bistratum. Proc. XXVII International Conference on Coordination Chemistry.* T23.

van den Brenk, A.L., Hanson, G.R. and Hawkins, C.J. (1991) Binding of copper II to a lipophilic cyclic octapeptide, patellamide D. (Manuscript in preparation).

Waisvicz, J.M., van der Hoeven, M.G. and Te Nijenhuis, B. (1957) The structure of the sulfur-containing moiety of bottromycin. *J. Am. Chem. Soc.* **79**, 4524.

Watters, D.J., Marshall, K., Hamilton, S., Michael, J., McArthur, M., Seymour, G., Hawkins, C., Gardiner, R. and Lavin, M.F. (1990) The bistratenes: New cytotoxic marine macrolides which induce some properties indicative of differentiation in HL-60 cells. *Biochem. Pharmacol.* **39**, 1609.

6 POISONING BY BLUE-GREEN ALGAE

IAN R. FALCONER

INTRODUCTION

Poisoning of livestock from drinking water containing blue-green algae (cyanobacteria) has been reported worldwide for over a century. The earliest detailed report was of stock deaths at Lake Alexandrina in South Australia in 1878 (Francis, 1978). The scum of blue-green algae was described as being like green oil paint, thick and pasty as porridge. The smell as it decomposed was likened to putrid urine. This same species occurs in the Baltic Sea, the Peel-Harvey estuary in Western Australia, and the East Gippsland Lakes. It is a filamentous organism, *Nodularia spumigena*, which thrives in brackish water.

Studies on the toxicity of this species have shown selective hepatotoxicity, the cause of death being either acute blood loss into the liver, or progressive liver failure (Runnegar *et al.*, 1988). Dogs are particularly vulnerable, as they accumulate the filaments in their coats when swimming and then lick themselves clean! Very sophisticated fast-atom bombardment mass spectroscopy has been used to identify the toxic agent, which is a cyclic pentapeptide shown in Figure 6.1 (Rinehart *et al.*, 1988). The molecular weight is 824 daltons, and the composition is a unique β-amino hydrophobic acid, γ-linked D-glutamic, N-methyl dehydrobutyrine, β-linked erythro β methyl D-aspartic, L-arginine. The toxin is called nodularin, it is resistant to boiling at neutral pH, and has a toxicity (LD_{50}) in the order of 70 μg/kg mice by intraperitoneal injection. The mechanism of toxicity appears to be through cytoskeletal deformation of hepatocytes leading to cell death and tissue disruption (Runnegar *et al.*, 1988).

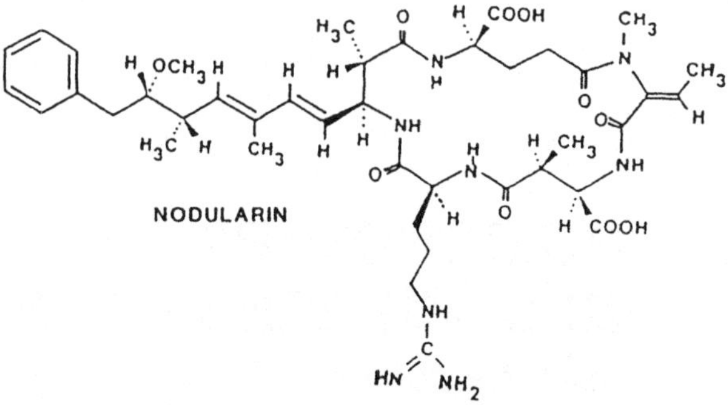

NODULARIN

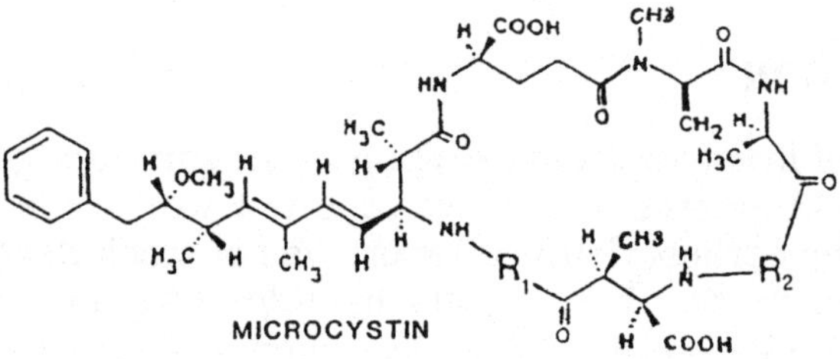

MICROCYSTIN

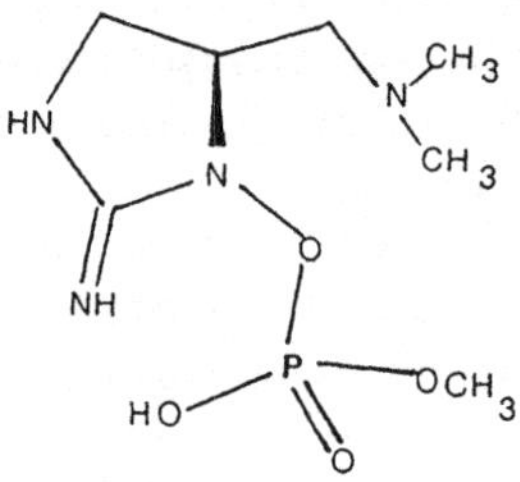

ANATOXIN-a(s)

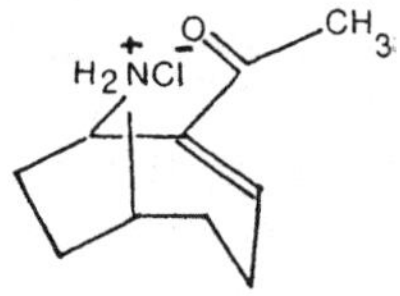

ANATOXIN-a

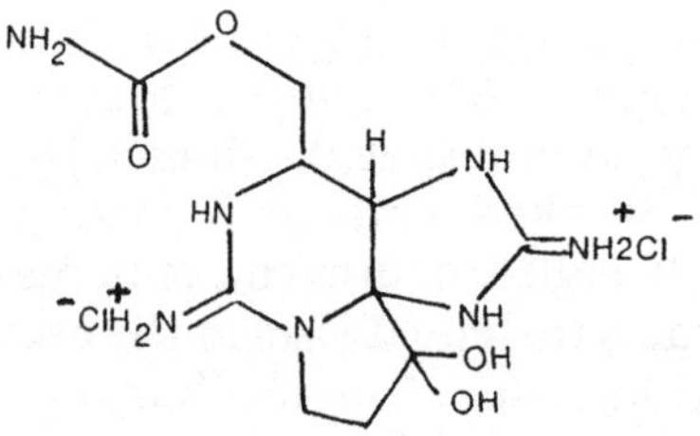

SAXITOXIN

Figure 6.1 Peptide and alkaloid toxins from cyanobacteria.

This cyanobacterium appears to be particularly responsive to the availability of phosphate in the water and sediments, and is capable of forming massive water-blooms in summer. The use of phosphatic fertilizers in the river catchments, and the high phosphate content of treated sewage effluents, contribute to the increasing incidence of these water-blooms (Lukatelich and McComb, 1986). Australian recreational uses of brackish water estuaries are likely to be severely affected by blue-green algae by the end of the century unless action is taken to reduce phosphate in streams and rivers.

In fresh-water lakes and reservoirs in Australia different species of cyanobacteria occur. One of these is *Microcystis aeruginosa*, which has caused livestock deaths on farms along the Dividing Range. It too forms water blooms, which when drunk by birds or mammals cause extensive liver damage and death. Because of the prevalence of this organism, and the formation of toxic blooms in the local city water supply, it has formed the main focus of our research.

The hepatotoxin, called microcystin, is one of a family of cyclic heptapeptides. Two L-amino acids in the ring are interchangeable without variation in toxicity, and one or several variant toxins may be obtained from the same isolate of cells. In the *Microcystis* toxin we have studied, the composition is similar to nodularin, the sequence being the same hydrophobic b amino acid, D-glutamic, N-methyl dehydroalanine, D-alanine, L-tyrosine, erythro β methyl D-aspartic, L-methionine (Figure 6.1) (Rinehart *et al.*, 1988). The toxicity (LD_{50}) is about 60 µg/kg mice by intraperitoneal injection. The acute toxic symptoms are massive structural disorganization in the liver and intrahepatic bleeding, with hepatocyte and sinusoidal disintegration. Sub-lethal doses cause hepatocyte necrosis, in the case of our toxin beginning in the peri-portal region of the liver lobules.

MECHANISMS OF TOXICITY

The mechanism of toxic action has been investigated in whole animals, tissue slices and isolated hepatocytes. There are no major changes in the metabolic function of affected cells, or in membrane permeability. In isolated hepatocytes the most obvious change is in cell morphology (Figure 6.2). Transmission electron microscopy of toxin-deformed cells shows aggregation of cytoplasmic elements just below bulbous projections from the cell

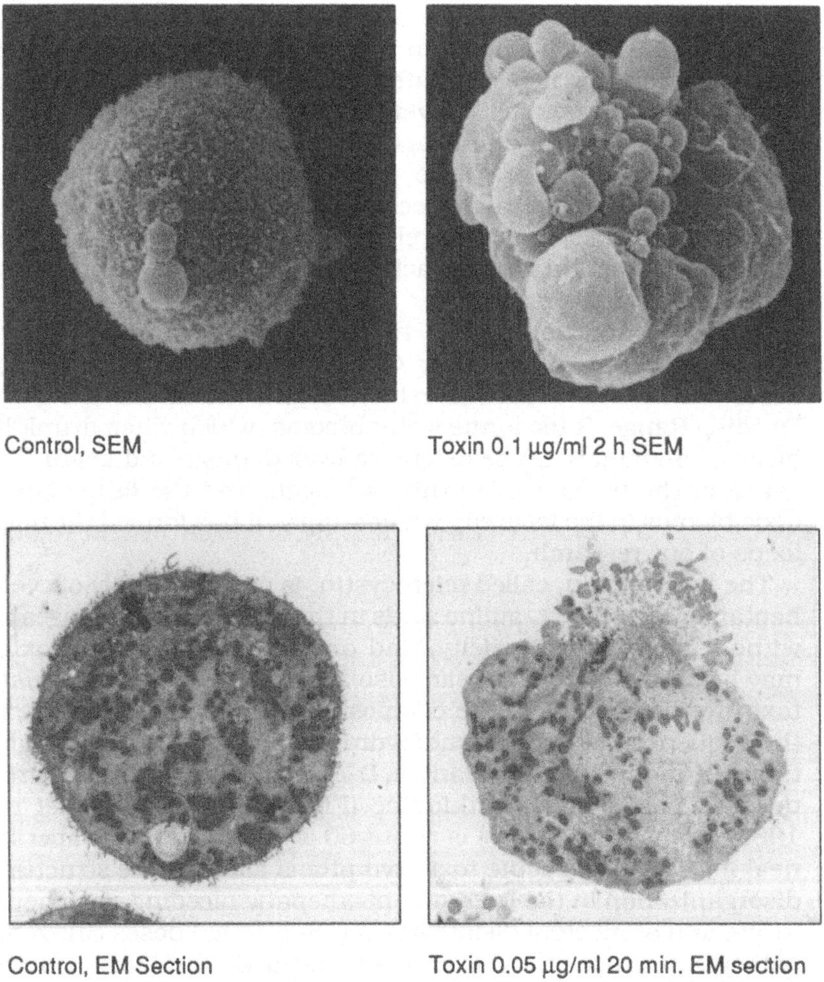

Figure 6.2 Morphological changes in hepatocytes incubated with microcystin. (Top right-hand figure reprinted with permission from *South African Journal of Science*, Volume 78, pp. 362–6, 1982.)

membrane. We are currently using antibodies to cytoskeletal proteins to try to visualize the composition of the deformed structure. Fluorescent-labelled second antibodies are used to identify the proteins concerned. The identity of the aggregated protein is also being explored by electrophoresis of cytoskeletal proteins from toxin-deformed hepatocytes. It is possible that covalent alterations in structural proteins are part of the toxic effect, since the deformation is only seen in living cells. In our attempts, no deformation of other cell types could be shown by incubation with toxin or by hypotonic lysis of red cells and re-sealing in the presence of toxin and ATP.

Cyanobacteria also synthesize and accumulate alkaloidal toxins. The common species *Anabaena circinalis*, which forms waterblooms in farm dams, has caused stock deaths in Eastern Australia. This organism contains neurotoxins, two of which have been fully identified (Figure 6.1). The heat-stable alkaloid anatoxin-a was found to be a blocking agent for postsynaptic neuromuscular transmission, acting through depolarization (Carmichael *et al.*, 1979). The clinical symptoms of poisoning in mice include leaping movements, abdominal breathing and convulsions occurring within a few minutes of an intraperitoneal dose. There are no apparent changes seen post-mortem. If the animal survives the acute neurotoxicity, apparent full recovery occurs within 3 hours.

This species of cyanobacterium has also been associated with allergic reactions in swimmers, which are probably not related to alkaloid content.

A second neurotoxic effect was observed in field cases of cyanobacterial poisoning, which exhibited quite different symptoms. In these occurrences the clinical signs include lacrimation, salivation, urination and diarrhoea. The toxin has been called anatoxin-a(s), and the structure just elucidated (Figure 6.1) (Mahmood and Carmichael, 1987; Matsunage *et al.*, 1989). The compound is a naturally-occurring organophosphate neurotoxin, related to guanidine. The toxicity is about 20 μg/kg in mice and is unstable to heat or alkalinity. It is an irreversible cholinesterase inhibitor.

The third type of neurotoxin is associated with a separate cyanobacterial genus, *Aphanizomenon*. This produces the sodium-channel blocking poison saxitoxin, which is better known as the cause of paralytic shellfish poisoning (Mahmood and Carmichael, 1986). So far this toxicity has only been identified in the USA.

The cyanobacteria are a very ancient and diverse group of organisms, which appear to have developed toxic mechanisms in the bloom-forming species to prevent predation. The toxic constituents can be assumed to be lethal or inhibitory to zooplankton, and where tested this has been supported experimentally. The toxic cyanobacteria provide a range of hazards to the human population, as well as to wild and domestic animals. Our laboratory is now particularly interested in chronic effects of ingestion of cyanobacteria, and has preliminary data for teratogenesis (Falconer *et al.*, 1988) and tumor promotion (Falconer and Buckley, 1989) in mice given continued *Microcystis* extracts in their drinking water. The public health implications of these findings are under investigation.

REFERENCES

Carmichael, W.W., Biggs, D.F. and Peterson, M.A. (1979) Pharmacology of anatoxin-a, produced by the freshwater cyanophyte *Anabaena flos-aquae* NRC-44-1. *Toxicon* **17**, 229–236.

Falconer, I.R. and Buckley, R.H. (1989) Tumour promotion by *Microcystis*, blue-green algae occurring in water supplies. *Med. J. Aust.* **150**, 351.

Falconer, I.R., Smith, J.V., Jackson, A.R.B., Jones, A. and Runnegar, M.T. (1988) Oral toxicity of a bloom of the cyanobacterium *Microcystis aeruginosa* administered to mice over periods of up to one year. *J. Toxicol. and Env. Health* **24**, 291–305.

Francis, G. (1878) Poisonous Australian lake. *Nature* **18**, 11–12.

Lukatelich, R.J. and McComb, A.J. (1986) Nutrient levels and the development of diatoms and blue-green algal blooms in a shallow Australian estuary. J. *Plankton Res.* **8**, 597–618.

Mahmood, N.A. and Carmichael, W.W. (1986) Paralytic shellfish poisons produced by the freshwater cyanobacterium *Aphanizomenon flos-aquae* NH-5. *Toxicon* **24**, 175–186.

Mahmood, N.A. and Carmichael, W.W. (1987) Anatoxin-a(s), an anticholinesterase from the cyanobacterium *Anabaena flos-aquae* NRC–525-17. *Toxicon* **25**, 1221–1227.

Matsunaga, S., Moore, R.E., Niemczura, W.P. and Carmichael, W.W. (1989) Anatoxin-a(s): a potent anticholinesterase from *Anabaena flos-aquae. J. Am. Chem. Soc.* **111**, 8021–8023.

Rinehart, K.L., Harada, K., Namikoshi, M., Chen, C. and Harvis, C.A. (1988) Nodularin, Microcystin, and the configuration of Adda. *J. Am. Chem. Soc.* **110**, 8557–8558.

Runnegar, M.T.C., Jackson, A.R.B. and Falconer, I.R. (1988) Toxicity of the cyanobacterium *Nodularia spumigena* Mertens. *Toxicon* **26**, 143–151.

7 SOFT CORALS AND THEIR TOXINS

BRUCE F. BOWDEN, PORFIRIO M. ALINO AND JOHN C. COLL

INTRODUCTION

The toxicity of the Zoanthids *Palythoa toxica*, and *Palythoa tuberculosa*, which contain palytoxin (probably the most toxic compound known from the marine environment, LD_{50} 0.03 μg/kg for dog) (Moore and Scheuer, 1971) has led to investigations of toxicity in the Alcyonacea. This interest has resulted in numerous reports of cytotoxicity, references to antifouling properties against filamentous algae (Coll *et al.*, 1987), barnacles (Keifer *et al.*, 1986) and diatoms (Bandurraga and Fenical, 1985; Targett *et al.*, 1985) as well as reports of insecticidal (Cardellina, 1986) and antimicrobial (Cardellina *et al.*, 1987) properties. Ichthyotoxicity of test samples against *Gambusia affinis* is a convenient bioassay which has been used in numerous studies by toxinologists. Ichthyotoxicity tests measure the toxicity of compounds (dissolved or suspended in water) which are presumably absorbed by the gill membranes of the fish. Previous work (Labarre *et al.*, 1986) has shown that approximately 50% of (>300) soft coral samples examined were toxic to *Gambusia affinis*. There was good correlation between lack of physical protection in the soft coral form and ichthyotoxicity, an observation which suggested chemical feeding deterrency may be important.

A number of species of soft corals have now been examined in detail and ranked according to their ichthyotoxicity towards *Gambusia affinis*. An extract of each coral was fractionated to yield a lipophilic, and a water-soluble fraction, and the relative toxicities of those fractions were determined. Solutions of some pure terpenoid compounds which were present in some of the corals used, were tested for ichthyotoxicity at various concentrations. Several of these compounds exhibited ichthyotoxicity at concentrations in the 5–10 ppm range. There appears to be some correlation between the relative toxicities of the lipophilic and water-soluble fractions, and the polarity of the terpenoid compounds known to be present in the corals.

Studies on a *Sarcophyton* species revealed that the concentration of the major diterpenoid metabolite (which was ichthyotoxic) was significantly higher in the actual coral polyp than in the surrounding coenenchyme tissue; although the presence of spicules alters the effective amount of tissue present in each case, significant concentration differences are still evident after correction for the weight of spicules present. This result may be significant for the coral in terms of protection against predation by Chaetodons (Pisces) which selectively browse on soft corals; indeed, selectivity is observed in the method of browsing by *Chaetodon melannotus* on specific soft corals, with coral polyps being selectively eaten for certain species, while the coenenchyme tissue and/or mucous of other species is preferred.

MATERIALS AND METHODS

Ichthyotoxicity

Gambusia affinis was exposed to crude aqueous extracts of nine octocorals (*Lobophytum* sp., *Sinularia flexibilis*, two *Sarcophyton* spp., *Pachyclavularia violacea*, *Clavularia inflata*, *Xenia* sp., *Alcyonium molle* and *Briareum steckei*) under conditions similar to those previously reported (La Barre *et al.*, 1986). This methodology was extended and modified as follows:

Each frozen octocoral sample (150 g) was placed in distilled water (300 ml) and macerated using a blender. The macerate was centrifuged at 10 000 rpm and the supernatant decanted. Three samples of this supernatant (10 ml) were used for ichthyotoxicity tests as replicate sets. Each test involved the use of three randomly picked *G. affinis* individuals collected from a nearby pond and placed in 200 ml of freshwater, in cubicle containers. Three replicate sets of fish in freshwater served as controls with a different batch of controls for every new set of trials. Fish mortality was recorded after 1.5 h, 3 h, 6 h, 12 h, and 24 h from exposure to the extract.

A further 100 ml of this supernatant was extracted with 100 ml of dichloromethane [DCM, = CH_2Cl_2] in a separatory funnel. This extraction was repeated three times with DCM in the usual way. The layers were allowed to separate, and the combined lower layers (300 ml), after evaporation on a vacuum rotary evaporator, served as the lipid soluble extract. The combined aqueous (upper) layers containing the water soluble components were also tested for ichthyotoxicity. The crude lipid soluble extract was dissolved in ethanol (10 ml) and 1 ml of the ethanol solution was added to another three replicate sets of *G. affinis*. Three sets of fish exposed to ethanol (1 ml) in conditioned freshwater, served as controls in these assays.

Ichthyotoxicity of Terpenes

Ten pure compounds which were some of the major terpenoid components from the nine octocoral species under investigation were also tested for ichthyotoxicity at the 5 ppm level using ethanol solution (1 mg/ml) diluted with conditioned freshwater (to 20 ml). The compounds tested (Figure 7.1) included: flexibilide (1) and dihydroflexibilide (2) from *Sinularia flexibilis*; (3) from *Lobophytum* sp., sarcophytoxide (4) from *Sarcophyton* sp.; a cladiellin (5) from *Alcyonium molle*; three cembranoids (6, 7, 8) from *Pachyclavularia violacea*; clavukerin (10) and a trisnorsesquiterpene (9) from *Clavularia inflata*; and a briaran derivative from *Briareum stechei* (11). Several compounds (3, 5, 7, 8, 9, 10, 11) were also tested at the 10 ppm level.

Comparison of Terpenoid Concentration in the Polyp and Coenenchyme of *Sarcophyton glaucum*

Samples were collected from Pioneer Bay, Orpheus Island (18°36'S & 146°29'30"E). Collection was made underwater by cutting off the extended polyps of *Sarcophyton glaucum* using scissors. The polyps tended to stick to the blades of the scissors, and were sucked up into a 10 ml plastic syringe (without needle). These samples were then transferred underwater into a sample vial, frozen and freeze-dried prior to analysis. Freeze-dried samples were weighed, extracted three times with DCM (1 ml), and the solvent evaporated using a rotary evaporator. The number of individual polyps were recorded in each sample to permit an estimation of the amount of terpene per individual polyp to be made. The crude DCM extracts containing mainly sarcophytoxide

(4) (MW = 306) were then weighed and dissolved in $CDCl_3$ containing a known amount of *m*-dinitrobenzene standard and a ^{1}H-NMR spectrum recorded. Proton peaks areas at d9.1, 8.6, 7.8 for m– dinitrobenzene and d5.5, 5.3, 5.1, 4.5 for sarcophytoxide were integrated. This afforded the required data for the computation of the amount of sarcophytoxide in each sample.

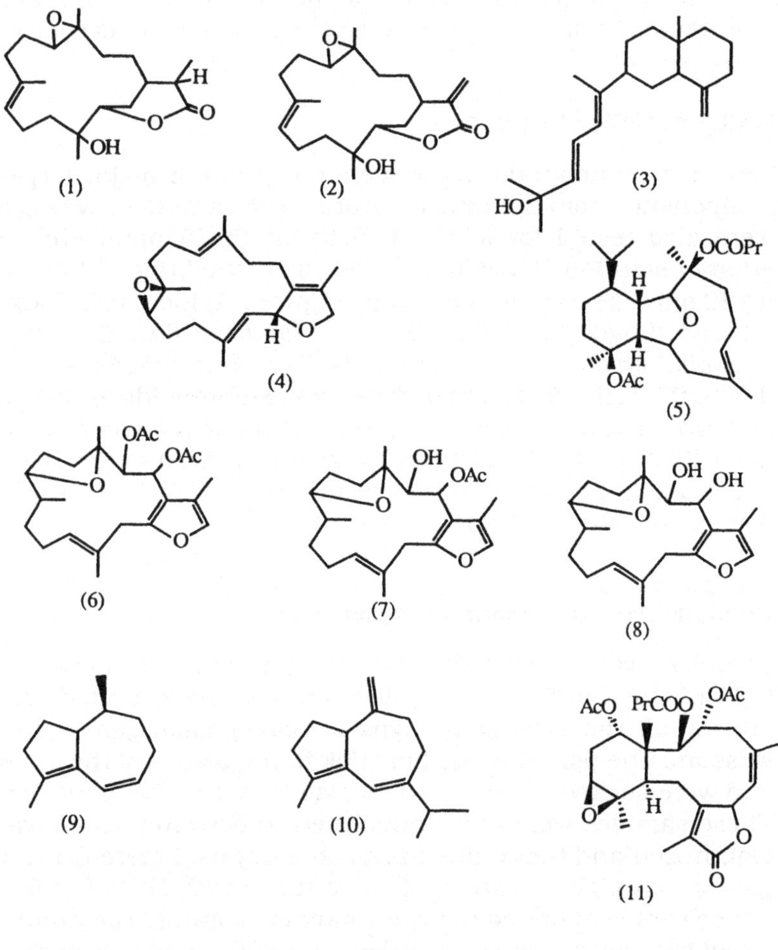

Figure 7.1. Structures of different compounds tested for ichthyotoxicity.

RESULTS

The results of the ichthyotoxicity tests of the unfractionated aqueous extracts of the nine octocorals in the survey are tabulated (Table 7.1).

Six of the octocorals were selected, and the crude extracts from each were fractionated to yield a lipophilic, and a water-soluble fraction. These fractions were similarly tested for ichthyotoxicity, and the results appear in Tables 7.2a and 7.2b.

Table 7.1 Summary of ichthyotoxicity tests on *G. affinis* using crude aqueous extracts of nine octocoral species

Species	*Mortality (proportion dead) Time in hours*					*Rank*
	1.5	*3*	*6*	*12*	*24*	
Lobophytum sp. FD5	1.00					(1)
Sinularia flexibilis FD3	0.77	1.00				(2)
Sarcophyton sp. FD6	0.77	1.00				(2)
Sarcophyton sp. FD8	0.22	1.00				(2)
Pachyclavularia violacea	0.11	1.00				(2)
Clavularia inflata	0.0	0.33	1.00			(3)
Xenia sp. FD3	0.0	0.33	0.88	1.00		(4)
Alcyonium molle	0.0	0.0	0.0	0.0	0.0	(5)
Briareum stechei	0.0	0.0	0.0	0.0	0.0	(5)
Controls pooled (n=12)	0.0	0.0	0.08	0.16	0.16	

Table 7.2a Summary of ichthyotoxicity tests on *G. affinis* using lipid soluble extracts of six octocoral species

Species	*Mortality (proportion dead) Time in hours*					*Rank*
	1.5	*3*	*6*	*12*	*24*	
Sinularia flexibilis FD2	0.00	0.00	0.22	0.44	1.00	(1)
Lobophytum sp. FD5	0.11	0.22	0.66	0.66	0.88	(2)
Pachyclavularia violacea	0.00	0.00	0.00	0.11	0.66	(3)
Clavularia inflata	0.00	0.00	0.00	0.00	0.44	(4)
Xenia sp. FD3	0.00	0.00	0.00	0.00	0.44	(4)
Briareum stechei	0.00	0.00	0.00	0.00	0.00	(5)
Control (EtOH)	0.0	0.0	0.00	0.00	0.08	

Table 7.2b Summary of ichthyotoxicity tests on *G. affinis* using water soluble extracts of six octocoral species

Species	*Mortality (proportion dead) Time in hours*					*Rank*
	1.5	*3*	*6*	*12*	*24*	
Sinularia flexibilis FD2	0.00	0.66	1.00			(1)
Xenia sp. FD3	0.00	0.33	0.44	0.66	0.66	(2)
Lobophytum sp. FD5	0.00	0.00	0.00	0.00	0.00	(3)
Briareum stechei	0.00	0.00	0.00	0.00	0.00	(3)
Pachyclavularia violacea	0.00	0.00	0.00	0.00	0.00	(3)
Clavularia inflata	0.00	0.00	0.00	0.00	0.00	(3)

Eleven pure terpenes were tested for ichthyotoxicity, and the results appear in Table 7.3.

The polyps of the *Sarcophyton* sp. were found to contain significantly higher concentrations of the terpene sarcophytoxide than were found in the coenenchyme tissues (Table 7.4).

DISCUSSION

From Table 7.1, a relative toxicity ranking and descriptive toxicity category: highly toxic (1,2), toxic (3,4), and non-toxic (5) can be assigned to each octocoral based on the initial toxicity tests.

The lipid-containing, DCM soluble fractions showed a similar trend in ichthyotoxicity (Table 7.2a) to the unfractionated aqueous extracts (Table 7.1). The ranking based on ichthyotoxicity of the water soluble extracts (Table 7.2b) however showed a shift in pattern, with a number of the formerly very toxic and highly toxic specimens becoming non-toxic.

Lipid soluble extracts		Water soluble extracts	
very toxic:	*S. flexibilis*	very toxic:	*S. flexibilis*
	Lobophytum sp.	toxic:	*Xenia* sp.
toxic:	*P. violacea*	nontoxic:	*B. stechei*
	Clavularia inflata		*Lobophytum* sp.
	Xenia sp.		*P. violacea*
non-toxic:	*B. stechei*		*C. inflata*

Table 7.3 Summary of ichthyotoxicity tests on *G. affinis* using some pure terpenoid compounds from octocorals

Compound			*Mortality (proportion dead) Time in hours*				
			1.5	*3*	*6*	*12*	*24*
P. violacea	5ppm	**(6)**	0.0	0.0	0.0	0.0	0.66
		(7)	0.0	0.0	0.0	0.0	0.88
		(8)	0.0	0.0	0.0	0.0	0.66
	10ppm	**(6)**	0.0	0.0	0.0	0.0	1.0
		(7)	0.0	0.0	0.0	0.11	1.0
		(8)	0.0	0.78	1.0	—	—
C. inflata	5ppm	**(9,10)**	0.0	0.0	0.0	0.0	0.0
	10ppm	**(9,10)**	0.0	0.0	0.22	0.56	0.67
B. stechei	5ppm	**(11)**	0.0	0.0	0.0	0.0	0.0
	10ppm	**(11)**	0.0	0.0	0.0	0.11	0.33
S. flexibilis		**(1)**	0.0	0.0	0.0	0.0	0.0
		(2)	0.0	0.0	0.0	0.0	0.22
Sarcophyton	5ppm	**(4)**	0.0	0.0	0.0	0.0	0.11
	10ppm	**(4)**	0.0	1.0	—	—	—
	(Tapiolas, 1985)						
L. pauciflorum	5ppm	**(3)**	0.0	0.0	0.0	0.00	0.11
A. molle	5ppm	**(5)**	0.0	0.0	0.0	0.11	0.33
	10ppm	**(5)**	0.0	0.0	0.0	0.0	1.0
Control without EtOH			0.0	0.0	0.0	0.07	0.14
Control with EtOH			0.0	0.0	0.0	0.0	0.03

It is notable that the terpenoid constituents in *S. flexibilis* and *Xenia* species are typically highly oxygenated and functionalized, and these compounds might be expected to have higher water solubilities than the more lipophilic terpenes. This fact could in part be responsible for the shift in ichthyotoxicity patterns observed for the water soluble extracts.

The ichthyotoxicity tests on *G. affinis* using pure terpenoid compounds from some specific octocorals indicated several compounds which were toxic at concentrations of 5–10 ppm. The

Figure 7.2. Examples of soft corals examined in this work, (above) *Clavularia inflata;* (below) *Dendronephthea.*

Table 7.4 Concentration of sarcophytoxide (KII) in the polyp and coenenchyme of *Sarcophyton glaucum*.

Coenenchyme			*Polyp*		
Extract (%, dry wt.)	*KII (%, total extract)*	*KII (%, dry wt.)*	*Extract (%, dry wt.)*	*KII (%, total extract)*	*KII (%, dry wt.)*
16.2	68.3	11.1	42.0	70.1	29.5
26.0	32.3	8.4	38.0	92.8	41.3
21.6	46.0	9.9	52.9	98.0	51.9
18.5	31.1	5.8	47.3	102.0*	50.2
25.2	43.9	11.1	45.5	74.2	26.5
21.5± 3.8	44.3± 13.4	9.2± 2.0	45.2±5.0	87.4±12.9	39.8±10.4

(Means with standard deviation)
* due to weighing errors.

three diterpene furans (6–8) from *Pachyclavularia violaceae* and the dihydrofuran (4) from *Sarcophyton* sp. were the most toxic, while the diterpene (5) from *Alcyonium molle* at a concentration of 10 ppm was apparently toxic, but much slower acting. The sesquiterpene hydrocarbons (9, 10) from *Clavularia inflata*, also appeared to have some activity.

Significantly higher concentrations of the diterpene sarcophytoxide (4) were found in the polyps of *Sarcophyton glaucum* than were found in the surrounding coenenchyme tissues. Approximately four times as much diterpene was present in the polyp than was detected in the surrounding tissues. This result was still valid after correction for the weight of spicules present.

REFERENCES

Bandurraga, M.M. and Fenical, W. (1985) Isolation of the muricins: evidence of a chemical adaptation against fouling in the marine ocotocoral *Muricea fructicosa* (Gorgonacea). *J. Chem. Ecol.* **12**, 951–987.

Cardellina, J.H. (1986) Marine natural products are leads to new pharmaceutical and agrochemical agents. *Pure and Appl. Chem.* **58**, 365–374.

Cardellina, J.H., Hendrickson, R.L. and Manfredi, K.P. (1987) Bissetone a unique antimicrobial pyranone from the gorgonian *Briareum polyanthes*. *Tetrahedron Letters* **28**, 727.

Coll, J.C., Price, I.R., Konig, G.M. and Bowden, B.F. (1987) Algal overgrowth of alcyonacean soft corals. *Mar. Biol.* **96**, 129–135.
Keifer, P.A., Rinehart, K.L. and Hooper, I.R. (1986) Renillafoulins, antifouling diterpenes from the sea pansy *Renilla reniformis* (Octocorallia). *J. Org. Chem.* **51**, 4450–54.
Labarre, S., Coll, J.C. and Sammarco, P.W. (1986) Defensive strategies of soft corals (Coelenterata: Octocorallia) of the Great Barrier Reef II. The relationship between toxicity and feeding deterrence. *Biol. Bull.* **171**, 565–574.
Moore, R.E. and Scheuer, P.J. (1971) Palytoxin: a new marine toxin from a coelenterate. *Science* **172**, 495–500.
Targett, N.M., Bishop, S.S., McConnell, O.J. and Yoder, J.A. (1985) Anti-fouling agents against the benthic marine diatom *Navicula salinocola*. Homarine from the gorgonian *Leptogorgia virgulata* and *Leptogorgia setacea* and analogs. *J. Chem. Ecol.* **9**, 817–830.

8 IMMUNOTHERAPY FOR CANCER—TOXIN-ANTIBODY CONJUGATES

GEOFFREY PIETERSZ AND IAN McKENZIE

INTRODUCTION

The response of most solid tumors to anticancer drugs is not satisfactory, and only a few tumors such as testicular tumor and childhood leukemias, can be cured whereas most other metastatic tumors continue to have a bad prognosis (Marsoni *et al.*, 1987). The main reason for this lack of efficacy is that most cytotoxic drugs cannot be used safely at their effective dose due to their low therapeutic index (Double *et al.*, 1989). By effectively targeting anticancer drugs to the tumor, most of the side effects resulting from damage to normal tissues can be reduced. This concept of the "Magic Bullet" was first proposed by Paul Ehrlich in the early 1900s (Ehrlich, 1913), but could not be effectively implemented due to the lack of specific antibodies. The discovery of tumor associated antigens and the production of monoclonal antibodies (MoAbs) to these antigens provided an impetus to re-examine the early concept of Paul Ehrlich (Kohler *et al.*, 1975; Hellstrom *et al.*, 1982). As a result, complexes obtained by linking drugs, toxins and radionuclides to MoAbs have been produced and these have been shown to selectively localize at the tumor and show superior antitumor effects to the non-conjugated moieties (Deland *et al.*, 1985; Youle *et al.*, 1986; Pietersz *et al.*, 1987). Most of the anticancer drugs used for immunoconjugation thus far have been those presently used in the clinic, and considering the inefficient uptake of the conjugate into tumor tissue due to physiological barriers, more toxic analogs would be more effective by decreasing the number of toxic molecules needed to kill a cell. This point can be illustrated when the antitumor efficacy of the more toxic Idarubicin and aminopterin immunoconjugates are compared to less toxic adriamycin and methotrexate immunoconjugates (Pietersz *et al.*, 1988). Idarubicin and aminopterin immunoconjugates are more effective antitumor agents *in vitro* and *in vivo* than their less toxic counterparts. Due to their extreme toxicity, toxins are attractive candidates for immunoconjugation (Table 8.1). These toxins are several thousand times more cytotoxic than commonly used antineoplastic agents. For example, it has been calculated that one ricin molecule entering a cell is sufficient to kill the cell (Eiklid *et al.*, 1980). The chemical methods for linking these toxins to monoclonal antibodies have been reviewed recently (Wawrzynczak *et al.*, 1987).

Table 8.1 Toxins conjugated to MoAb

Toxin		*Reference*
Plant holotoxins	Ricin	
	whole	Weil-Hillman *et al.*, 1985
	A-chain	Blakey *et al.*, 1988
	Abrin	
	whole	Thorpe *et al.*, 1982
	A-chain	Thorpe *et al.*, 1987
Plant hemitoxins	Gelonin	Reimann *et al.*, 1988
	Saporin	Bregni *et al.*, 1988
	Poke weed antiviral protein	Masuho *et al.*, 1982
	Trichokirin	Casellas *et al.*, 1988
	Bryodin	Stirpe *et al.*, 1988
	Mormordin	Stirpe *et al.*, 1988
Animal	Sea anemone hemolytic toxin	Avila *et al.*, 1988
	Cobra venom factor	Petrella *et al.*, 1987
Fungal	α-Amanitin	Davis *et al.*, 1981
	Restrictocin	Conde *et al.*, 1989
Bacterial	Diphtheria	Bumol *et al.*, 1983
	Pseudomonas exotoxin	FitzGerald *et al.*, 1986

RICIN IMMUNOTOXINS

Ricin is a glycoprotein isolated from the seeds of the castor bean plant *Ricinus communis* and has been one of the most studied toxins for coupling to monoclonal antibodies. Ricin has a molecular weight of 65 kD consisting of two disulphide-linked chains of molecular weight 32 kD (A-chain) and 33 kD (B-chain) (Olsnes *et al.*, 1973). The B-chain has a galactose binding domain and is an essential part of the molecule for internalization and translocation of the A-chain into the cytoplasm. The A-chain inhibits eukaryotic protein synthesis by cleaving a single N-glycosidic bond at A-4324 of the 28S rRNA in the 60S ribosome (Endo *et al.*, 1987). Several strategies have been described to link ricin to MoAbs. Linking the A-chain of ricin to MoAb via a disulphide bond results in a highly selective immunotoxin where the MoAb mimics the B-chain function by binding to cell surface antigens. However, lack of natural B-chain, thought to be involved in transport of the A-chain from the endocytic vesicle, results in inefficient translocation of the A-chain to the cytosol and A-chain immunotoxins are occasionally inactive or less active than the native toxin. However, these A-chain immunotoxins may be potentiated by using ionophores, free B-chain or lysosomotropic agents such as ammonium chloride (McIntosh *et al.*, 1983; Ramakrishnan *et al.*, 1989). The linkage of intact toxin to MoAb can result in conjugates that display carbohydrate binding of the B-chain in addition to antibody binding, and these conjugates will show the antibody specificity only when used in the presence of lactose. These intact ricin immunotoxins have been useful for *ex vivo* purging of bone marrow for transplantation (Vallera *et al.*, 1987) but would have limited *in vivo* usefulness. Another method of obtaining specific intact ricin antibody conjugates is to use ricin which has been blocked at the galactose binding site by chemical modification. In our studies, linkage of ricin to antibody using the cross-linking agents SPDP and SAMSA resulted in a disulphide-linked intact ricin immunotoxin devoid of galactose binding, probably due to steric obstruction of the galactose binding site by the MoAb (Pietersz *et al.*, 1988). These ricin-anti-Ly-2.1 conjugates were selectively cytotoxic to cells expressing the Ly-2.1 antigen and moreover they were as cytotoxic as unconjugated whole ricin. Indeed, these are the most potent immunoconjugates described to date. The ricin-anti-Ly-2.1 conjugate inhibited protein synthesis of Ly-2.1^{+ve}, Ly-1.1^{-ve} cells by 50% at 45 ng/ml whereas non-reactive ricin-anti-Ly-1.1 conjugate was ineffective. When mice bearing an intraperitoneal Ly-2.1^{+ve} thymoma were treated intraperitoneally (i.p.) with ricin-anti-Ly-2.1, 90% of the mice survived indefinitely, whereas all control (saline-treated) mice, died by day 14 (Table 8.2) (Kanellos *et al.*, 1988). Ricin-anti-Ly-2.1 conjugates given intravenously were also capable of decreasing the growth of subcutaneous tumors in mice (Table 8.2). When these immunotoxins were administered directly into the subcutaneous tumor, within 24 h all tumors were eradicated (Kanellos *et al.*, 1989) (Table 8.2). Similar efficacy have been shown when mice bearing human colon tumors were directly injected with ricin-anti-human colon tumor antibodies. Thus, if a highly toxic immunoconjugate is put in the right place in the right amount, it can eradicate a solid tumor. On the basis of these results, Phase I clinical trials are currently in progress for the treatment of breast and colon cancer, by intravenous or regional administration of ricin conjugates.

GENETIC ENGINEERING OF TOXINS AND IMMUNOTOXINS

In addition to plant holotoxins, several other toxins of microbial origin have functional domains for binding, translocation and toxicity. Several early experiments have shown that binding domains from one toxin may be exchanged with a catalytic domain of another toxin, resulting in a functional hybrid toxin (Sundan *et al.*, 1982). These biochemical manipulations can also be performed using genetic engineering techniques. The most notable work has been performed using the microbial toxins— diphtheria toxin (DT) and *Pseudomonas* exotoxin (PE). Both these toxins have a domain with ADP-ribosylating activity in addition to separate domains involved in binding and translocation. Cell death

Table 8.2 Efficacy of anti-Ly-2.1-ricin immunotoxin on a Ly-2^{+ve} thymoma

Site of tumor	*Mode of delivery*	*Dose of ricin-anti-Ly-2.1 (μg of ricin)*	*% Control tumor size on day 16*
i.p	i.p	0.05 μg (day 1)	> 200 days*
s.c	i.v.	0.1 μg × 2 (day 8 and 11)	30%
s.c.	i.t	0.2 mg × 2 (day 7 and 9)	10%

* 90% of mice survived >200 days, while control animals died on day 55.

results after toxin internalization due to the ADP-ribosylation of elongation factor-2 involved in protein synthesis (Collier, 1975; Hudson *et al.*, 1987). Using recombinant DNA techniques, the binding domain of PE has been deleted, to form non-toxic PE-40 which has been engineered to produce fusion proteins with TGF-a, IL-2, IL-4, IL-6 and CD4 (Chaudhary *et al.*, 1987, 1988; Lorberboum-Galski *et al.*, 1988; Siegall *et al.*, 1988; Ogata *et al.*, 1989) (Table 8.3). CRM-45, a mutant of DT, lacking the binding domain, has also been fused to IL-2 (William *et al.*, 1989). These fusion proteins show exquisite selectivity to cells expressing the receptor and are more active than the equivalent chemically linked conjugates. They can be used as antitumor agents, for immunosuppression in transplantation, arthritis and AIDS (Kelley *et al.*, 1988; Kirkman *et al.*, 1989; Pankerwycz *et al.*, 1989). The ricin gene has also been cloned as the preproricin precursor (Lamb *et al.*, 1985), (post-translational modifications leading to the mature form of ricin). In addition, biologically active A-chain and B-chain of ricin have been expressed in bacteria and yeast (Hussain *et al.*, 1989). The crystal structure of ricin has also been determined and the amino acids involved in galactose binding are known. Using techniques such as site-directed mutagenesis, B-chains devoid of galactose binding may now be produced (Monfort *et al.*, 1987). These mutant B-chains may be combined with recombinant A-chains to form intact ricin which could be used for immunoconjugation or the mutant B-chains could be used to potentiate A-chain immunotoxins *in vivo* or used for the production of piggyback toxins. Immunoconjugates of recombinant A-chain have already been used in clinical trials, and these recombinant toxins will also be more favoured than the native material due to the lack of carbohydrate, which is present in the

Table 8.3 Recombinant immunotoxins

Toxin fragment	*Receptor binding domain*	*Reference*
PE-40	IL-2	Lorberboum-Galski *et al.*, 1988
	IL-4	Ogata *et al.*, 1989
	IL-6	Siegall *et al.*, 1988
	TGF-α	Chaudhary *et al.*, 1987
	Fv (anti-IL-2R)	Chaudhary *et al.*, 1989
	CD4	Chaudhary *et al.*, 1988
CRM-45	IL-2	William *et al.*, 1987

native toxin causing it to bind to liver and cause toxic effects (Thorpe *et al.*, 1985).

PROBLEMS WITH TOXINS

The use of toxins is accompanied by several problems, such as toxicity—discussed above, their size and their immunogenicity. Size could be a problem, as coupling ricin and antibody could result in a substantial increase in molecular weight of the complex which may have some difficulty in penetration of tumors. Ricin A-chain, as opposed to ricin may partly overcome this problem. In addition, toxin-antibody conjugates are likely to be immunogenic and lead to both anti-toxin and anti-antibody responses. The latter may be overcome by "humanizing" antibodies, but anti-toxin responses are likely to interfere with the potency of toxin-antibody conjugates. Recently single chain molecules functioning as the Fv fragment of antibody have been genetically engineered (Huston *et al.*, 1988). These fragments consist of a part of the light chain and heavy chain hypervariable regions linked by a peptide linker molecule. These fragments with their small molecular weight could form versatile carriers for drugs or toxins. An elegant extension to this technology is the engineering of a fusion protein consisting of the Fv fragment of an antibody to the human IL-2 receptor and PE-40 (Chaudhary *et al.*, 1989). This immunotoxin with its smaller molecular weight (65 kD) may be more efficient in permeating tumors than the larger chemically synthesized immunotoxins (215 kD).

CONCLUSION

Monoclonal antibodies when linked to toxins result in specific immunotoxins that can be used to kill populations of cells that react with the antibody. One of the disadvantages of these immunotoxins is their size, and as a result would not efficiently permeate into the tumor and highly potent toxins of lower molecular weight may be more efficient for killing tumors. Using recombinant DNA technology these small highly potent toxins may be selectively manipulated and unwanted properties eradicated.

Thus far, there have been several clinical trials using either whole ricin-MoAb or A-chain-MoAb conjugates and some therapeutic effects noted. It remains to be seen whether genetically engineered materials can lead to a further improvement in the treatment of cancer.

REFERENCES

Avila, A.D., Mateo de Acosta, C. and Lage, A. (1988) A new immunotoxin built by linking a haemolytic toxin to a monoclonal antibody specific for immature T lymphocytes. *Int. J. Cancer* **42**, 568–571.

Blakey, D.A. and Thorpe, P.E. (1988) An overview of therapy with immunotoxins containing ricin or its A-chain. *Antibody Immunoconjugates and Radiopharmaceuticals* **1**, 1–16.

Bregni, M., Lappi, D.A., Siena, S., Formosa, A., Villa, S., Sonia, M., Bonadonna, G. and Masimo Gianni, A. (1988) Activity of a monoclonal antibody-saporin-6 conjugate against B-lymphoma cells. *J. Natl. Cancer Inst.* **80**, 511–517.

Bumol, T.F., Wang, Q.C., Reisfeld, R.A. and Kaplan, N.O. (1983) Monoclonal antibody and an antibody toxin conjugate to a cell surface proteoglycan of melanoma cells suppress *in vivo* tumor growth. *Proc. Natl. Acad. Sci. USA* **80**, 529–533.

Casellas, P., Bourrie, B.J.P., Gros, P. and Jansen, F.K. (1984) Kinetics of cytotoxicity induced by immunotoxins: Enhancement by lysosomotropic amines and carboxylic ionophores. *J. Biol. Chem.* **259**, 9359–9364.

Chaudhary, V.K., Fitzgerald, D.J., Adhya, S. and Pastan, I. (1987) Activity of a recombinant fusion protein between transforming growth factor type a and *Pseudomonas* toxin. *Proc. Natl. Acad. Sci. USA* **84**, 1538–1542.

Chaudhary, V.K., Mizukami, T., Fuerst, T.R., Fitzgerald, D.J., Moss, B., Pastan, I. and Berger, E.A. (1988) Selective killing of HIV-infected cells by recombinant CD4-*Pseudomonas* exotoxin hybrid protein. *Nature* **335**, 369–372.

Chaudhary, V.K., Queen, C., Junghan, R.P., Waldmann, T.A., Fitzgerald, D.J. and Pastan, I. (1989) A recombinant immunotoxin consisting of two antibody variable domains fused to *Pseudomonas* exotoxin. *Nature* **339**, 394–397.

Collier, R.J. (1975) Diphtheria toxin: mode of action and structure. *Bacteriol. Rev.* **39**, 54–85.

Conde, F.P., Orlandi, R., Canevari, S., Mezzanzanica, D., Ripamonti, M., Munoz, S.M., Jorge, P. and Colnaghi, M.I. (1989) The *Aspergillus* toxin restrictocin is a suitable cytotoxic agent for generation of immunoconjugates with monoclonal antibodies directed against human carcinoma cells. *Eur. J. Biochem.* **178**, 795–802.

Davis, M-T.B. and Preston, J.F. (1981) A conjugate of a-amanitin and monoclonal immunoglobulin G to Thy 1.2 antigen is selectively toxic to T lymphoma cells. *Science* **213**, 1385–1388.

Deland, F.H. and Goldenberg, D.M. (1985) Diagnosis and treatment of neoplasms with radionuclide– labelled antibodies. *Semin. Nucl. Med.* **15**, 2–11.

Double, J.A. and Bibby, M.C. (1989) Therapeutic index: A vital component in selection of anticancer agents for clinical trial. *J. Natl. Cancer Inst.* **81**, 988–994.

Ehrlich, P. (1913) Chemotherapy, Proc. 17th Int. Congress Med. In: *The Collected Papers of Ehrlich*, edited by F. Himmelweit, Vol. III, p. 510, London: Pergamon Press.

Eiklid, K., Olsnes, S. and Pihl, A. (1980) Entry of lethal doses of abrin, ricin and modeccin into Hela cells. *Exp. Cell Res.* **126**, 321–326.

Endo, Y., Mitsui, K., Motizuki, M. *et al.* (1987) The mechanism of action of ricin and related toxic lectins on eukaryotic ribosomes. *J. Biol. Chem.* **262**, 5908–5912.

Fitzgerald, D.J., Willingham, M.C. and Pastan, I. (1986) Antitumor effects of an immunotoxin made with *Pseudomonas* exotoxin in a nude mouse model of human ovarian cancer. *Proc. Natl. Acad. Sci. USA* **83**, 6627–6630.

Hellström, K.E., Hellström, I., Brown, J.P. (1982) Human tumor associated antigens identified by monoclonal antibodies. *Springer Semin. Immunopathol.* **5**, 127–146.

Hudson, T.H. and Neville, Jr, D.M. (1987) Temporal separation of protein toxin translocation from processing events. *J. Biol. Chem.* **262**, 16484–16494.

Hussain, K., Bowler, C., Roberts, L.M. and Lord, J.M. (1989) Expression of ricin B-chain in *Escherichia coli*. *FEBS Lett.* **244**, 383–387.

Huston, J.S., Levinson, D., Mudgett-Hunter, M., Tai, M-S., Novotny, J., Margolies, M.N., Ridge, R.J., Bruccoleri, R.E., Haber, E., Crea, R. and Oppermann, H. (1988) Protein engineering of antibody binding sites. Recovery of specific activity is an anti-digoxin single-chain Fv analogue produced in *Escherichia coli*. *Proc. Natl. Acad. Sci. USA* **85**, 5879–5883.

Kanellos, J., McKenzie, I.F.C. and Pietersz, G.A. (1988) *In vivo* studies of whole ricin monoclonal antibody immunoconjugates for the treatment of murine tumors. *Immunol. Cell Biol.* **66**, 403–415.

Kanellos, J., McKenzie, I.F.C. and Pietersz, G.A. (1989) Intratumor therapy of solid tumors with ricin-antibody conjugates. *Immunol. Cell Biol.* **67**, 89–99.

Kelley, V.E., Bacha, P., Pankewycz, O., Nichols, J.C., Murphy, J.R. and Strom, T.B. (1988) Interleukin 2-diphtheria toxin fusion protein can abolish cell mediated immunity *in vivo*. *Proc. Natl. Acad. Sci. USA* **85**, 3980–3984.

Kirkman, R.L., Bacha, P., Barrett, L.V., Forte, S., Murphy, J.R. and Strom, T.B. (1989) Prolongation of cardiac allograft survival in murine recipients treated with a diphtheria toxin-related interleukin-2 fusion protein. *Transplantation* **47**, 327–330.

Köhler, G. and Milstein, C. (1975) Continuous cultures of fused cells secreting antibody of predefined specificity. *Nature* **256**, 495–497.

Lamb, F.I., Roberts, L.M. and Lord, J.M. (1985) Nucleotide sequence of cloned cDNA coding for preproricin. *Eur. J. Biochem.* **148**, 265–270.

Lorberboum-Galski, H., Fitzgerald, D.J.P., Chaudhary, V., Adhya, S. and Pastan, I. (1988) Cytotoxic activity of an interleukin 2-Pseudomonas exotoxin chimeric protein produced in *Escherichia coli*. *Proc. Natl. Acad. Sci. USA* **85**, 1922–1926.

McIntosh, D.P., Edwards, D.C., Cumber, A.J., Parnell, G.D., Dean, C.J., Ross, W.C.J. and Forrester, J.A. (1983) Ricin B-chain converts a non-cytotoxic antibody-ricin conjugate into a potent and specific cytotoxic agent. *FEBS Lett.* **164**, 17–20.

Marsoni, S., Hoth, D., Simon, R., *et al.* (1987) Clinical drug development. An analysis of phase II trials, 1970–1985. *Cancer Treat. Rep.* **71**, 71–80.

Masuho, Y., Kishida, K. and Hara, T. (1982) Targeting of the antiviral protein from *Phytolacca americana* with an antibody. *Biochem. Biophys. Res. Comm.* **105**, 462–469.

Montford, W., Villafranca, J.E., Manzingo, A.F., Ernst, S.R., Katsin, B., Rutenber, E., Young, N.H., Hamlin, R. and Robertus, J.S. (1987) The three-dimensional structure of ricin at 2.8 A. *J. Biol. Chem.* **252**, 5398–5403.

Ogata, M., Chaudhary, V.K., Fitzgerald, D.J. and Pastan, I. (1989) Cytotoxic activity of a recombinant fusion protein between interleukin 4 and *Pseudomonas* exotoxin. *Proc. Natl. Acad. Sci. USA* **86**, 4215–4219.

Olsnes, S. and Pihl, A. (1973) Different biological properties of the two constituent peptide chains of ricin, a toxic protein inhibiting protein synthesis. *Biochem.* **12**, 3121–3126.

Pankerwycz, O., Mackie, J., Hassarjain, R., Murphy, J.R., Strom, T.B. and Kelley, V.E. (1989) Interleukin-2-diphtheria toxin fusion protein prolongs murine islet cell on engraftment. *Transplantation* **47**, 318–322.

Petrella, E.C., Wilkie, S.D., Smith, C.A., Morgan, Jr, A.C. and Vogel, C-W. (1987) Antibody conjugates with cobra venom factor. *J. Immunol. Methods* **104**, 159–172.

Pietersz, G.A., Kanellos, J. and McKenzie, I.F.C. (1988) Novel synthesis and *in vitro* characterization of disulfide-linked ricin-monoclonal antibody conjugates devoid of galactose binding activity. *Cancer Res.* **48**, 4469–4476.

Pietersz, G.A., Kanellos, J., Smyth, M.J., Zalcberg, J. and McKenzie, I.F.C. (1987) The use of monoclonal antibody conjugates for the diagnosis and treatment of cancer. *Immunol. Cell Biol.* **65(2)**, 111–125.

Pietersz, G.A., Smyth, M.J. and McKenzie, I.F.C. (1988) Immunochemotherapy of a murine thymoma with the use of Idarubicin monoclonal antibody conjugates. *Cancer Res.* **48**, 926–931.

Ramakrishnan, S., Bjorn, M.J. and Houston, L.L. (1989) Recombinant ricin A-chain conjugated to monoclonal antibodies. Improved tumor cell inhibition in the presence of lysomotropic compounds. *Cancer Res.* **49**, 613–617.

Reimann, K.A., Goldmacher, V.S., Lambert, J.M., Chalifoux, L.V., Cook, S.B., Schlossman, S.F. and Letvin, N.L. (1988) *In vivo* administration of lymphocyte specific monoclonal antibodies in non-human primates. IV. Cytotoxic effect of an anti-T11-gelonin immunotoxin. *J. Clin. Invest.* **82**, 129–138.

Siegall, C.B., Chaudhary, V.K., Fitzgerald, D.J. and Pastan, I. (1988) Cytotoxic activity of an interleukin - 6 - Pseudomonas exotoxin fusion protein on human myeloma cells. *Proc. Natl. Acad. Sci. USA* **85**, 9738–9742.

Stirpe, M.F., Wawrzynczak, E.J., Brown, E.J., Knyba, R.E., Watson, G-J., Barbieri, L. and Thorpe, P.E. (1988) Selective cytotoxic activity of immunotoxins composed of a monoclonal anti-Thy 1.1 antibody and the ribosome inactivating proteins bryodin and mormordin. *Br. J. Cancer* **58**, 558–561.

Sundan, A., Olsnes, S., Sandvig, K. and Pihl, A. (1982) Preparation and properties of chimeric toxins prepared from the constituent polypeptides of diphtheria toxin and ricin. *J. Biol. Chem.* **259**, 9733–9739.

Thorpe, P.E., Blakey, D.C., Brown, A.N.F., Knowles, P.P., Knyba, R.E., Wallace, P.M., Watson, G.J. and Wawrzynczak, E.J. (1987) Comparison of two anti-Thy-1.1 Abrin A-chain immunotoxins prepared with different cross-linking agents: antitumor effects, *in vivo* fate, and tumor cell mutants. *J. Natl. Cancer Inst.* **79**, 1101–1112.

Thorpe, P.E., Detre, S.I., Foxwell, B.M.J. *et al.* (1985) Modification of the carbohydrate in ricin with metaperiodate-cyanoborohydride mixtures. Effects on toxicity and *in vivo* distribution. *Eur. J. Biochem.* **147**, 197–206.

Thorpe, P.E. and Ross, W.C.J. (1982) The preparation and cytotoxic properties of antibody-toxin conjugates. *Immunol. Rev.* **62**, 119–158.

Vallera, D.A. (1987). The use of immunotoxins in bone marrow transplantation: Eradication of T cells and leukemic cells. In: *Immunoconjugates: Antibody Conjugates in Radioimaging and Therapy of Cancer*, edited by C.-W. Vogel, pp. 217–240. New York: Oxford University Press.

Wawrzynczak, E.J. and Thorpe, P.E. (1987) Methods for preparing immunotoxins. Effect of the linkage on activity and stability. In: *Immunoconjugates: Antibody Conjugates in Radioimaging and Therapy of Cancer*, edited by C.-W. Vogel, pp. 28-55. New York: Oxford University Press.

Weil-Hillman, G., Runge, W., Jansen, F.K. and Vallera, D.A. (1985) Cytotoxic effect of anti-Mr 67 000 protein immunotoxins on human tumours in a nude mouse model. *Cancer Res.* **45**, 1328–1336.

William, D.P., Parker, K., Bacha, P. *et al.* (1987) Diphtheria toxin receptor binding domain substitution with interleukin-2: genetic construction and properties of a diphtheria toxin-related interleukin-2 fusion protein. *Protein Engineering* **1**, 493.

Youle, R.J. and Colombatti, M. (1986) Immunotoxins: monoclonal antibodies linked to toxic proteins for bone marrow transplantation and cancer therapy. In: *Monoclonal Antibodies in Cancer. Advances in Diagnosis and treatment*, edited by J.A. Roth, pp. 173–213. Mt Kisco, NY: Futura Publishing Co.

9 A TICK TOXIN

MICHAEL J. THURN AND KEVIN W. BROADY

INTRODUCTION

Thousands of animals die annually along the eastern seaboard of Australia as a result of the paralysis tick, *Ixodes holocyclus* (see Figure 9.1). The cause is attributed to a toxin produced in the tick's salivary glands during engorgement (Binnington and Stone, 1981). Toxicosis in dogs progresses from vomiting to rapid ascending paralysis of the limbs. If untreated, death results from respiratory paralysis. Current treatment is expensive and involves the use of an undefined antiserum.

Little is known about the toxin. It has yet to be characterized and its mode and site of action elucidated. To this end, the toxin is being purified and specific antibodies developed. Earlier attempts at purifying the toxin (Kaire, 1966 and Stone, 1979) were hampered by techniques which were too crude to provide substantial purification. Current techniques (HPLC and immuno-affinity chromatography) are much more specific and allow substances present in minute amounts to be purified.

METHODS

Crude Tick Preparation: 100 frozen ticks (engorged on mice for 3–5 days) were homogenized in a 2 ml Wheaton micro tissue grinder containing ice-cold distilled water (1 tick/100 ml) for 10 min. The insoluble material was then removed by centrifugation at 10 000 g for 10 min and 4°C.

Toxin Assay: The crude extract and the isolated fractions were injected (100 ml) subcutaneously into 5 day old (4–5 g) QS mice. Mice were observed for characteristic paralysis after 16 hours.

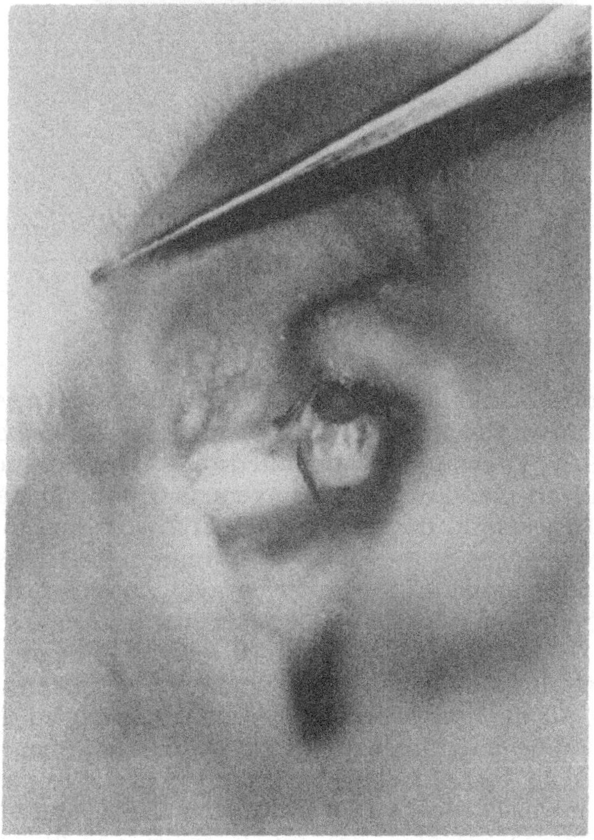

Figure 9.1 A female adult paralysis tick (*Ixodes holocyclus*) engorging on the ear of a mouse (approx. three days after attachment).

Molecular Sieve Chromatography: Crude extracts (0.5 ml) were layered onto a 1 × 28 cm column of AcA 34 Ultrogel (LKB) equilibrated with 0.1 M ammonium acetate pH 7.1 at room temperature. The column was eluted with the same buffer at a flow rate of 7.5 ml/h and fractions (7.5 min) collected (see Figure 9.2).

Anion Exchange Chromatography: The toxic fraction from the AcA 34 column was chromatographed on a 1.5 × 7 cm column of DEAE Trisacryl M (IBF) equilibrated with 0.01 M ammonium acetate pH 7.1. The column was eluted stepwise with increasing ionic strength ammonium acetate at a flow rate of 7.0 ml/h and fractions (7.0 min) collected (see Figure 9.3).

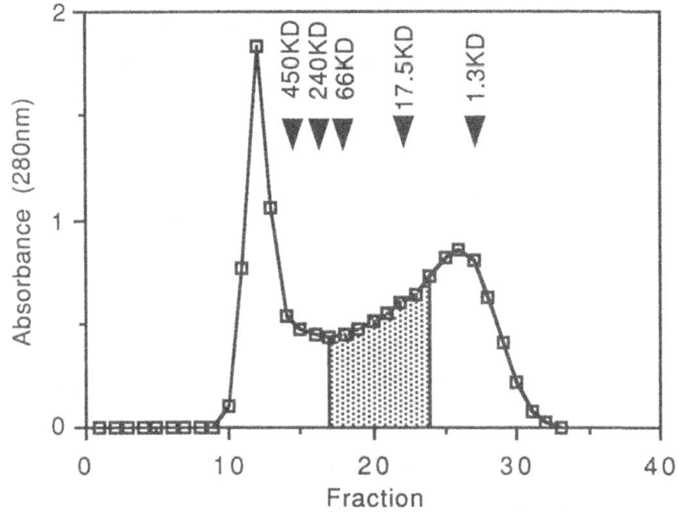

Figure 9.2 Molecular sieve chromatography. Elution profile of crude tick preparation on AcA 34 Ultrogel. The toxic fractions are indicated by shading. Molecular weight standards; 450 kD, ferritin; 240 kD, catalase; 66 kD, bovine serum albumin; 17.5 kD, myoglobin; 1.3 kD, vitamin B_{12}.

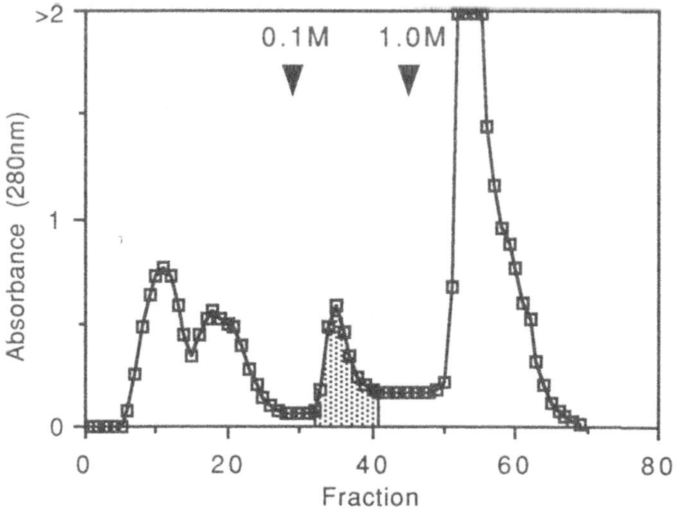

Figure 9.3 Anion exchange chromatography. Elution profile of the pooled toxic fragments from the molecular sieve chromatography on DEAE Trisacryl M. The toxic fractions are indicated by shading. The increases in the ionic strength of the eluting ammonium acetate buffer are indicated.

HPLC Ion Exchange Chromatography: Gradient elution high performance liquid chromatography (EPT Kortec system) was carried out with a Mono Q HR5/5 (Pharmacia) column, at a flow rate of 1.0 ml/min. Ammonium acetate 1.0 M, pH 7.1, was used as the buffer and the eluant was collected with a FRAC-100 fraction collector at 1.0 min intervals (see Figure 9.4).

SDS PAGE: The crude extract preparation and the various isolated fractions were resolved under reducing conditions by sodium dodecyl sulphate (SDS) polyacrylamide gel electrophoresis (PAGE) using a discontinuous gel (73 × 83 × 0.75 mm) system (Laemmli, 1970) with an acrylamide concentration of 12% in the separating gel. Gels were fixed and stained with Coomassie brilliant blue and Silver nitrate (see Figure 9.5).

Protein Determination: Protein was determined by the Coomassie brilliant blue G-250 dye-binding assay (Bradford, 1976). Bovine serum albumin (Sigma) was used as the standard.

RESULTS

Molecular sieve chromatography was employed to remove high molecular weight material from the crude extract and to furnish a toxic fraction suitable for anion exchange chromatography. The resulting toxic fraction had an LD_{50} of 20.5 mg/kg while the crude extract had an LD_{50} of 47.5 mg/kg.

Further purification by anion exchange chromatography gave a toxic fraction eluting with low ionic strength buffer. The 0.1 M toxic fraction had an LD_{50} of 3 mg/kg. Subsequent purification by HPLC produced a 0.2 M toxic fraction with an LD_{50} of 2 mg/kg.

DISCUSSION

The principle aim of this study was to produce a purified extract of tick toxin suitable for monoclonal antibody production. The toxic fraction isolated by anion exchange chromatography is one such example. Once a monoclonal antibody is produced, the final purification of the toxin by immuno-affinity chromatography can commence. This will allow the toxin to be produced in sufficient amounts, facilitating biochemical, pharmacological, and immunological studies of the toxin.

Until then, further purification of the HPLC toxic fraction showing a prominent band at an apparent molecular weight of 66 kD by SDS-PAGE, will be by gradient elution HPLC.

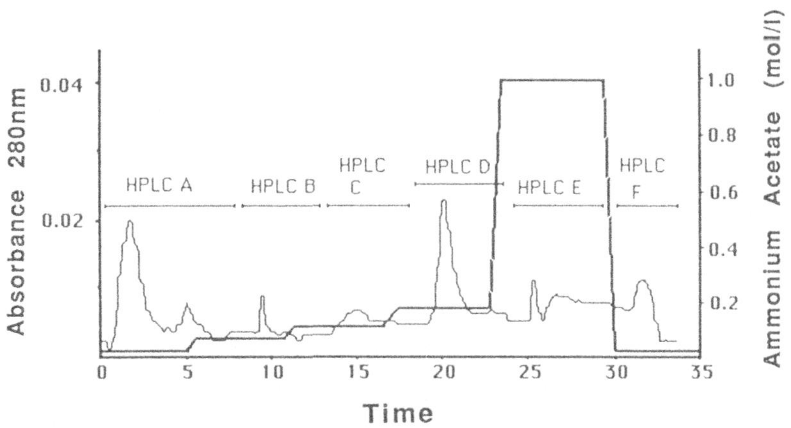

Figure 9.4 HPLC anion exchange chromatography. Elution profile of the pooled toxic fractions from anion exchange chromatography on a Mono Q HR5/5 column. Toxic activity was found in pooled fraction D.

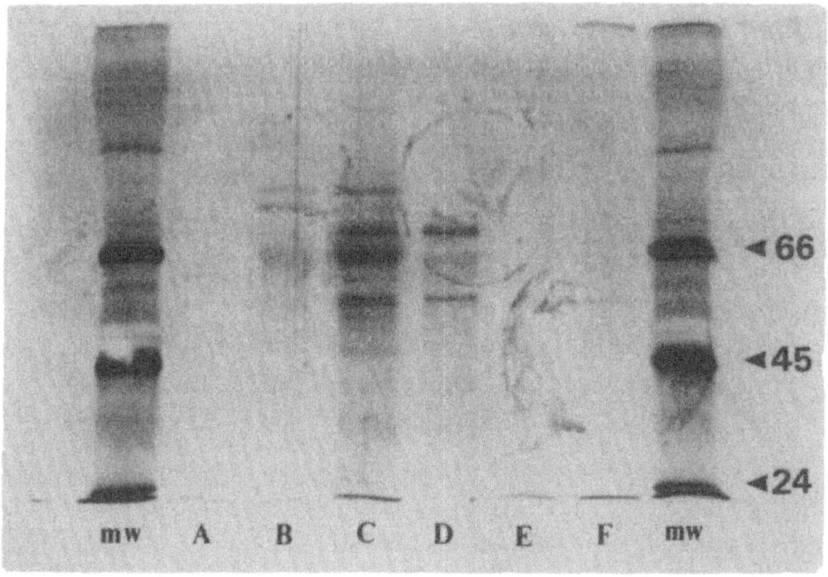

Figure 9.5 SDS-PAGE of HPLC fractions. Molecular weight markers: bovine albumin, 66 kD; trypsinogen, 24 kD. A,B,C,D,E and F are the respective pooled fractions from HPLC anion exchange chromatography. The gel has been silver stained.

REFERENCES

Binnington, K.C. and Stone, B.F. (1981) Developmental changes in morphology and toxin content of the salivary gland of the Australian paralysis tick *Ixodes holocyclus*. *Int. J. Parasitol.* **11**, 343–51.

Bradford, M.M. (1976) A rapid and sensitive method for the quantification of microgram quantities of protein, utilizing the principle of protein-dye-binding. *Anal. Biochem.* **72**, 248–54.

Kaire, G.H. (1966) Isolation of tick paralysis toxin from *Ixodes holocyclus*. *Toxicon* **4**, 91–97.

Laemmli, U.K. (1970) Cleavage of structural proteins during the assembly of the head of Bacteriophage T4. *Nature* **227**, 680–85.

Stone, B.F. (1979) Chemical characterization studies on salivary gland toxins of the paralysis tick *Ixodes holocyclus*. In: *Neurotoxins: Fundamental and Clinical Advances*, edited by Chubb, I.W. and Geffen, L.B., p. 273, Adelaide: Adelaide University, Union Press.

10 A CYCAD NEUROTOXIN

ALAN A. SEAWRIGHT, ALWYN W. BROWN, CHRISTOPHER C. NOLAN AND JOHN B. CAVANAGH

INTRODUCTION

The cycads are a group of relict Mesozoic plants of which 10 genera now exist in the mainly tropical to warm temperate regions of the world. Four of these genera are represented in Australia, namely *Cycas* (10 species), *Macrozamia* (14 species), *Lepidozamia* (two species) and *Bowenia* (two species) (Johnson, 1959). The mature fruits (Figure 10.1) and other parts of these plants have been known to be toxic when eaten by man and animals from

Figure 10.1 Mature fruit of a cycad.

time immemorial. The main toxic effect in mammals is acute liver damage which is due to methylazoxymethanol (MAM) glycosides, cycasins and macrozamins (Hegnauer, 1986).

An additional toxicity which affects cattle after weeks of consumption of cycad leaves is a posterior ataxia known locally as "zamia staggers" (Hall and McGavin, 1968; Hooper *et al.*, 1974). A toxin responsible for this condition has not so far been discovered. Since 1954 when Kurland and Mulder suggested that long term consumption of cycad starch or medicines by local inhabitants could be the cause of amyotrophic lateral sclerosis (ALS) and other dementia-motor neurone diseases in the Guam native population, there has been an interest in the possible neurotoxicity for man of products derived from these plants. The first report of a neurotoxin isolated from the plants was that of α-amino-β-methylaminopropionic acid (L-BMAA) from *Cycas circinalis* by Vega and Bell in 1967. This non-protein amino acid was of low toxicity for mice, rats and chickens (Polsky *et al.*, 1972), as well as being present in low concentration in the plants. It was thought at the time to be of little or no significance in relation to either "zamia staggers" of cattle or ALS in man.

Interest in the compound was revived by Spencer *et al.* (1987) recently when it was shown that a daily oral intake of the compound over several weeks led to an ALS-like disease in monkeys.

Although in the original acute toxicity studies with the compound in chickens, mice and rats, a neurotoxic effect was observed, no neuropathological changes were reported. Lately the acute toxic effects of this compound in young rats was confirmed and specific anatomical changes in the brain recorded.

MATERIALS AND METHODS

The compound (L-BMAA) was synthesized in the hydrochloride form as described by Vega *et al.* (1968). The crystals were dissolved in 1N NaOH so as to give a neutral solution and injected intraperitoneally into young male and female Wistar rats of 80 g body weight at from 6 to 14 mmoles/g body weight. At varying periods after dosing the animals were anesthetized with ether and their brains perfused-fixed with FAM for light microscopy or with Karnovsky's fixative for light and electron microscopy (Seawright *et al.*, 1990).

RESULTS AND DISCUSSION

About one hour after injection the rats became unsteady and by two hours began to rock from side to side. By six hours, when startled, they went into a slow torsion spasm and began rolling over and over. By 24 hours they were very ataxic, making padding movements across the floor with the tail arched over the body and nose on the floor. Rolling, spinning motions could easily be elicited. After 48 hours the ataxia was worse and they could not walk without falling. Animals given the higher doses were killed for necropsy at this stage while those given lower doses tended to be completely recovered within a further two to three days. During the period of the neurological disturbances body weights fell markedly but weight gains occurred once more with apparent recovery.

No gross pathological changes were observed in any organs. Microscopically, lesions were seen only in the nervous system and these were confined to the cerebellum. In animals killed at 6–7 hours after dosing with the L-BMAA, only a few pyknotic nuclei were to be seen in the molecular layer of the cerebellar cortex. This effect was more marked by 24 hours by which time the necrosis was widespread among the stellate and basket neurones of mainly the anterior vermis. With the higher doses, the cortex of the lateral lobes was also involved. After this Purkinje cells in the most severely affected areas were also affected and there was vacuolation of the neuropile of the cerebellar roof nuclei and also in the granular layer of the cortex. In animals kept for as long as five days there were areas of gliosis in the affected cerebellar cortex and regional increases in Bergmann glial nuclei.

Electron microscopic changes were also apparent at six hours and apart from degeneration of neuronal cytoplasmic components in affected stellate cells, there was swelling of Purkinje cell dendrites which contained masses of myelin figure-like material. These changes were more marked by 24 hours. By 48 hours cytoplasmic degeneration was noted in some Purkinje cells as well, and cell necrosis was frequent. Fragments of damaged Purkinje cell dendrites could be seen in the molecular layer together with electron dense amorphous masses derived from the earlier formed myelin figures. Granule cells remained normal throughout.

The acute neurological effects in the rats were similar to those earlier described by Polsky *et al.* (1972). These were typical disturbances of both posture and movement as would have been expected from damage to cerebellar neurones. There appeared to be an initial effect on the stellate, and basket neurones of the cerebeller cortex and this was followed by damage to Purkinje cells and the changes in the cerebellar roof nuclei. The changes seen in the electron microscope were typical of those due to excitotoxic action (Olney, 1980). The feature common to all the damaged neurones is that they were all GABAergic while the undamaged granule cell neurones use glutamate as a neurotransmitter.

Evidence exists both *in vitro* and *in vivo* that the effect of L-BMAA on mouse brain can be antagonized by the NMDA antagonist 2-amino-7-phosphoheptanoic acid but not by other glutamate antagonists (Ross *et al.*, 1987; Ross and Spencer, 1987). Moreover, it has been shown that L-BMAA, after interaction with bicarbonate becomes potentially excitotoxic via NMDA—preferring receptors. This suggests that the action of L-BMAA *in vivo* in the rat cerebellar cortex is the result of NMDA agonist activity. Rat Purkinje cells appear to have NMDA receptors but they do not react in the same way to NMDA antagonists. It can be presumed therefore that the effects on these cells are secondary to the primary excitotoxic effects on the stellate—basket neurones which project on to them (Figure 10.2). These suggestions remain to be further investigated and the basis of the regional susceptibility of the various glutamate receptor neurones in the rat cerebellar cortex requires further study.

These lesions in the rat brain due to a single dose of L-BMAA administered intraperitoneally in no way appear to resemble the peripheral sensory ataxia of "zamia staggers" in cattle nor the ALS-like disease of monkeys given multiple oral doses of L-BMAA. L-BMAA is apparently found only in *Cycas* sp and has not been isolated from other cycad genera which also cause "zamia staggers" both in Australia and elsewhere (Mason and Whiting, 1968, Kobayashi *et al.*, 1983). Also the content of L-BMAA in the plant (about 30 mg/100 g green leaf) would give a maximum daily intake of the compound for a calf eating cycad leaves of about one thousandth of that given the rats in the present study. The compound may be of relevance to ALS in Guam and other areas in which *Cycas* sp are growing but seem unlikely to be involved in the cause of "zamia staggers" of livestock.

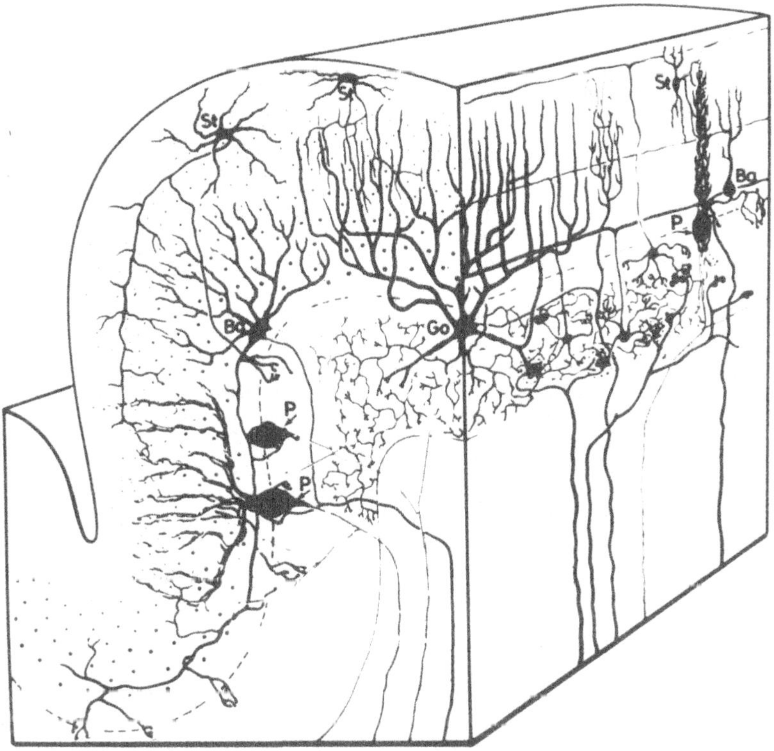

Figure 10.2 Neuronal connectivities in the cerebellum. Ba - Basket cell, Go - Golgi cell, P – Purkinje cell, St – Stellate cell. Reprinted with permission from *The Cerebellum as a Neuronal Machine*. Copyright 1967, Springer-Verlag.

REFERENCES

Hall, W.T.K. and McGavin, M.D. (1968) Clinical and neuropathological changes in cattle eating the leaves of *Macrozamia lucida* and *Bowenia serrulata* (Family Zamiaceae). *Path. Vet.* **5**, 26.

Hegnauer, R. (1986) Cycadopsida. In: *Chemotaxonomie der Pflanzen* **7**, 469. Stuttgart: Birkhaüser Verlag.

Hooper, P.T., Best, S.M. and Campbell, A. (1974) Axonal dystrophy in the spinal cords of cattle consuming the cycad palm, *Cycas media*. *Aust. Vet. J.* **50**, 146.

Johnson, L.A.S. (1959) The families of cycads and the Zamiaceae of Australia. *Proc. Linn. Soc. N.S.W.* **84**, 64.

Kobayashi, A., Todera, K., Yagi, F., Kono, I. and Yoshuda, N. (1983) Cattle poisoning due to ingestion of cycad leaves. Neurotoxic effects causing paralysis in hindquarters. *Toxicon. suppl.* **3**, 229.

Kurland, L.T. and Mulder, D.W. (1954) Epidemiologic investigations of amyotrophic lateral sclerosis. *Neurology* **4**, 355.

Mason, M.M. and Whiting, M.G. (1968) Caudal motor weakness and ataxia in cattle in the Caribbean area following ingestion of cycads. *Cornell Vet.* **58**, 541.

Olney, J.W. (1980) Excitotoxic mechanisms of neurotoxicity. In: *Experimental and Clinical Neurotoxicology*, edited by P.S. Spencer and H.H. Schaumburg, p. 175. Baltimore: Williams and Wilkins.

Polsky, F.I., Nunn, P.B. and Bell, E.A. (1972) Distribution and toxicity of α-amino-β-methylaminopropionic acid. *Procs. Fedn. Am. Socs. Exp. Biol.* **31**, 1473.

Ross, S.M., Seelig, M. and Spencer, P.S. (1987) Specific antagonism of excitotoxic action of "uncommon" amino acids assayed in organotypic mouse cortical cultures. *Brain Research* **425**, 120.

Ross, S.M. and Spencer, P.S. (1987) Specific antagonism of behavioural action of "uncommon" amino acids linked to motor system diseases. *Synapse* **1**, 248.

Seawright, A.A., Brown, A.W., Nolan, C.C. and Cavanagh, J.B. (1990) Selective degeneration of cerebellar cortical neurones caused by cycad neurotoxin, L-BMAA in rats. *Neuropath. Appl. Neurobiol.* **16**, 153–169.

Spencer, P.S., Nunn, P.B., Hugon, J., Ludolph, A.C., Ross, S.M., Roy, D.N. and Robertson, R.C. (1987) Guam amyotrophic lateral sclerosis–Parkinsonism–Dementia linked to a plant excitant neurotoxin. *Science* **237**, 517.

Vega, A. and Bell, E.A. (1967) α-Amino-β-methylaminopropionic acid, a new amino acid from seeds of *Cycas circinalis*. *Phytochemistry* **6**, 759.

Vega, A., Bell, E.A. and Nunn, P.B. (1968) The preparation of L- and D-α-amino-β-methylaminopropionic acids and the identification of the compound isolated from *Cycas circinalis* as the L-isomer. *Phytochemistry* **7**, 1885.

11 CIGUATERA POISONING

MICHAEL F. CAPRA AND JOHN CAMERON

INTRODUCTION

Each year, in Australia, a small but significant number of people will develop a complex of symptoms after eating certain tropical fish. Most will experience an acute episode of illness for a few weeks while some will develop chronic symptoms which may persist for months or even years. This illness, known as ciguatera poisoning, is not confined to Australia and has been shown to have a circumtropical distribution with a close association to coral reefs.

The ciguatera syndrome was documented by Europeans in the West Indies as early as the sixteenth century (Withers, 1982). A description of ciguatera poisoning in the Pacific was given by Anderson (1776). Anderson, the surgeon's mate on the *Resolution* during Cook's second voyage of discovery to the South Seas, gave a detailed account of an incident in which several of Cook's officers were poisoned after eating fish from the waters of Vanuatu. The symptoms described by Anderson closely match those that can be found in any modern review of ciguatera poisoning. Anderson also noted that the only treatment available was " to palliate the symptoms from the analogy they bore to those which occur in other diseases". Over 200 years have passed since Anderson's account, yet the treatment of ciguatera poisoning is still essentially symptomatic.

Ciguatera poisoning is currently a considerable cause of morbidity in tropical regions with an estimated 50 000 people being afflicted each year (Ragelis, 1984). Ciguatera poisoning often becomes a significant and serious health problem in the small island nations of the tropics where fish can form a considerable portion of the diet. The Caribbean and the Pacific contain many small island states where ciguatera poisoning has become a local health problem (Lewis, 1984; Taylor, 1984). In more affluent countries, ciguatera poisoning, in addition to being a health hazard, could become detrimental to large sections of the fishing industry. In the USA, ciguatera poisoning is the most frequently reported form of food-borne disease of a chemical nature (Withers, 1982). A recent successful litigation (Maran, 1990) following a case of ciguatera poisoning in America, may lead to a decreased tendency for hotels and restaurants to serve tropical fish. Proceedings are still before New South Wales courts after a substantial outbreak of ciguatera poisoning in Sydney in 1987.

SIGNS AND SYMPTOMS OF CIGUATERA POISONING

The clinical syndrome of ciguatera poisoning has been well documented by many authors over the years (Bagnis *et al.*, 1979; Lawrence *et al.*, 1980; Gillespie *et al.*, 1986; Bagnis and Legrand, 1987; Cameron and Capra, 1990). Following ingestion of toxic fish, the course of the syndrome follows a reasonably predictable pattern. Initial symptoms are usually gastrointestinal and develop 3 to 12 hours after a meal of fish; the severity of these symptoms is variable and may depend upon the toxicity and quantity of ingested fish. Gastrointestinal symptoms may include nausea, vomiting, diarrhoea and abdominal cramps. Neurological symptoms usually begin to appear after the onset of gastrointestinal symptoms. The neurological symptoms may include paraesthesia, arthralgia, myalgia, dental pain, convulsions, muscular paralysis, audio and visual hallucinations, vertigo, severe headache, diaphoresis, loss of short term memory and temperature perception reversals. Some patients have shown psychological disturbances, manifested as anxiety states and depression, for several months after intoxication. Skin rashes on the limbs, neck and trunk often occur within a few days to a few weeks of the ingestion of toxic fish. Cardiovascular symptoms including bradycardia, tachycardia, arrhythmias and hypotension are also noted in a number of victims.

The neurological disturbance in ciguatera poisoning may vary from mild discomfort over a period of a few days to more severe symptoms that can persist for months (Lawrence *et al.*, 1980) or even years (Capra and Cameron, 1985). The ingestion of alcohol can exacerbate the symptoms and can, in some victims, re-establish symptoms after an apparent recovery from an episode of poisoning.

Although the symptoms of ciguatera poisoning are often severe and the course of the illness can be prolonged, the incidence of mortality is low. A comprehensive study of some 3000 victims of ciguatera poisoning in the Pacific by Bagnis *et al.*, (1979) revealed a mortality rate of less than 0.1%. In Australia, there has been only one well documented death attributed to a severe CTX intoxication (Tonge *et al.*, 1967).

ORIGINS OF CIGUATERA POISONING

Human ciguatera intoxication can follow the consumption of a variety of tropical fish. In Queensland three species of fish (Red Bass, Chinaman Fish and Paddle-tail) are widely recognized as being potentially poisonous and cannot be legally sold. Capra (1985) assessed the toxicity of these species captured from reefs within a 60 km radius of Cairns and estimated that the incidences of toxicity were approximately 4% in Red Bass, 7% in Chinaman Fish and 18% in paddle-tail. On the basis of these figures a sufficiently high incidence of toxicity occurs to preclude the human consumption of these fish from this locality.

In addition to the species that are known to be high risk fish, a number of other species can at certain time and in particular localities, produce the ciguatera syndrome. Spanish Mackerel, Coral Trout and Barracuda are some of the commercially important species that have caused human poisoning in Australia.

A number of episodes of multiple ciguatera poisonings have occurred in Australia after the sale of toxic fish. The largest single episode occurred in Sydney in 1987 when 63 people became ill after eating Spanish Mackerel bought from a variety of retail outlets. The outbreak was eventually traced to a single large wholesaler in Sydney. Initial information suggested that the offending fish had originated in northern New South Wales. Subsequent investigation, however, indicated that the fish was captured in a ciguatera endemic area of Hervey Bay in Queensland and passed through the Queensland Fish Markets enroute to Sydney. The implications of the Sydney outbreak may yet affect the Queensland fishing industry as court proceedings against the wholesaler are now before the New South Wales supreme court. If the wholesaler is held responsible for selling toxic fish there is likely to be a marked decline in the sale of fish from any potentially toxic region.

Structural elucidation of ciguatoxin, the principal molecule believed to be responsible for the ciguatera syndrome, has proved to be a difficult task. The general properties and molecular weight of the ciguatoxin molecule (MW 1111) have been known since the early 1980s (Tachibana, 1980) when it was established that ciguatoxin was a polycyclic ether with extremely high toxicity (LD_{50} 0.45 mg/kg mice).

The first full structure for ciguatoxin from eels was published in 1989 by Murata *et al.* The symptomatology of ciguatera poisoning is not always constant with individual symptoms or groups of symptoms being absent or exaggerated in individual victims. Such variation may reflect individual differences in susceptibility and/or the quantity of toxin ingested. Another possibility that cannot be excluded is that the ciguatera syndrome may reflect exposure to a variety of related toxins rather than a single entity. Kodama *et al.*, 1989 have recently suggested that other marine toxins including palytoxin may contribute to the ciguatera syndrome.

The origins of ciguatoxin remained obscure until the diets of carrier species were closely examined in endemically toxic areas of French Polynesia (Yasumoto *et al.*, 1977a). Large numbers of a dinoflagellate *Diplopsalis* sp. were found in a toxic fraction of detritus from the Gambier Islands and extracts from these dinoflagellate rich samples had similar chromatographic, pharmacological and immunological properties to CTX derived from fish tissues (Yasumoto *et al.*, 1977b). The dinoflagellate, initially classified as *Diplopsalis*, was subsequently described as *Gambierdiscus toxicus* (Adachi and Fukuyo, 1979). *Gambierdiscus*, an epiphyte found on a number of species of macroalgae, has now been identified in Hawaii (Taylor, 1979), the Caribbean (Miller *et al.*, 1982), areas of the Pacific other than French Polynesia (Yasumoto *et al.*, 1984) and Australia (Gillespie *et al.*, 1985).

EFFECTS OF CIGUATOXIN ON MAMMALIAN NERVES

Despite many clinical reports describing the neurological signs and symptoms of ciguatera poisoning, very little has been documented as to the electrophysiological disturbance ciguatoxin causes in the peripheral nervous system. Over the past six years our research group has investigated the effects of ciguatoxin extracts from toxic fish on the electrical properties of the mammalian peripheral nervous system.

Ciguatoxin is believed to act at the molecular level on voltage dependent Na^+ channels to increase the permeability of excitable membranes to Na^+ (Bidard *et al.*, 1984). The action of ciguatoxin on Na^+ channels is believed to be responsible for the alterations observed in a number of nerve conduction parameters in isolated preparations (Boyarsky and Rayner, 1970; Lewis and Endean, 1983). While isolated preparations are useful in obtaining fundamental information, the responses of isolated nerves may not correspond to the responses of nerves within their natural environment. Capra and Cameron (1985) used an *in vivo* technique of recording nerve activity in rats to investigate the actions of ciguatoxin on nerve conduction. This technique has been used by our group over the past five years to provide an insight into the mode of action of ciguatoxin on the nerves of intact animals. Essentially, the ventral coccygeal nerve of an anesthetized rat is stimulated distally by surface electrodes and induced nerve action potentials are recorded proximally from another set of surface electrodes. One particular nerve conduction parameter, supernormality, has proved to be most useful in our study of the mode of action of ciguatoxin. For a very brief period (1–2 ms) after a nerve carries an action potential, a period of refractoriness occurs in which it is impossible or very difficult to activate Na^+ channels and thus elicit a second nerve action potential. After the refractory period the Na^+ gating mechanism becomes less stable and for a period of 10–30 ms it is easier to open the Na^+ gates and initiate a nerve action potential. This period is known as the supernormal period. Supernormality can be measured from compound nerves such as the rat ventral coccygeal nerve by giving a test stimulus that in isolation produces about a one-third maximal response. This test stimulus can then be preceded at variable time intervals by a supramaximal conditioning stimulus. The ratio of the conditioned test stimulus to the unconditioned test stimulus can be used to plot supernormality.

When a sublethal dose of ciguatoxin is given to a rat there is a significant increase in both the amplitude and duration of the supernormal period (Cameron, Flowers and Capra, 1991a) (Figure 11.1). Supernormality is related to the lability and ease of opening of voltage dependent Na^+ gates, hence any compound that increases the opening ease or the time that Na^+ gates reopen might be expected to enhance the magnitude of supernormality

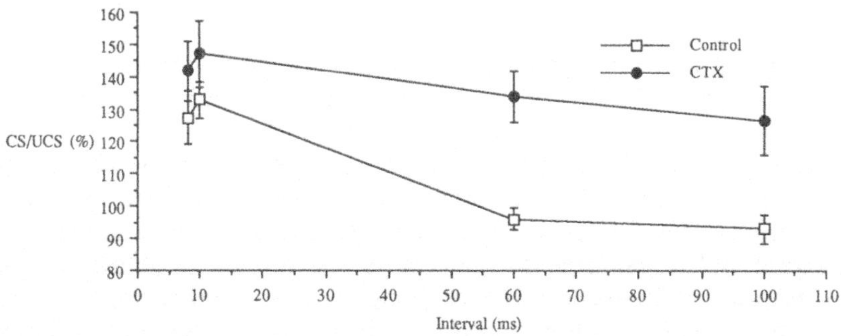

Figure 11.1 Percentage change in amplitude (conditioned stimulus (CS)/unconditioned stimulus (UCS)) of response to submaximal stimulus when preceded by a supramaximal conditioning stimulus at interstimulus intervals of 8, 10, 60 and 100 ms. Open boxes represent the means of 22 control rats and closed circles 21 rats treated with ciguatoxins. Error bars represent the standard errors of the means. (Reproduced with permission from the *Journal of Neurological Sciences* **101**, 87–92, 1991, Elsevier Science Publishers.)

and increase the duration of the supernormal period. The ciguatoxin that our group has extracted from fish would appear to be such a compound.

In 1987 two outbreaks of human ciguatera poisoning occurred in Australia– one in Sydney N.S.W. and the other in Maryborough, Queensland. Supernormal studies of the Sural nerve conduction were performed on a total of 15 victims demonstrating acute symptoms and signs of ciguatera poisoning (seven from Sydney and eight from Maryborough) (Cameron, Flowers and Capra, 1991b). All patients gave informal consent for these studies and their ages ranged from 19 to 57 (mean 32). Control studies were performed on 15 age-matched non-poisoned individuals with no obvious neurological dysfunction.

Figure 11.2 shows that both the magnitude and duration of the supernormal period are significantly increased in the victims of ciguatera poisoning when compared to a non-poisoned control group.

EFFECTS OF CIGUATOXIN ON FISH NERVES

The fact that a single fish can carry sufficient ciguatoxin to poison several humans yet shows no overt signs of intoxication has always intrigued our group. A number of studies have now been

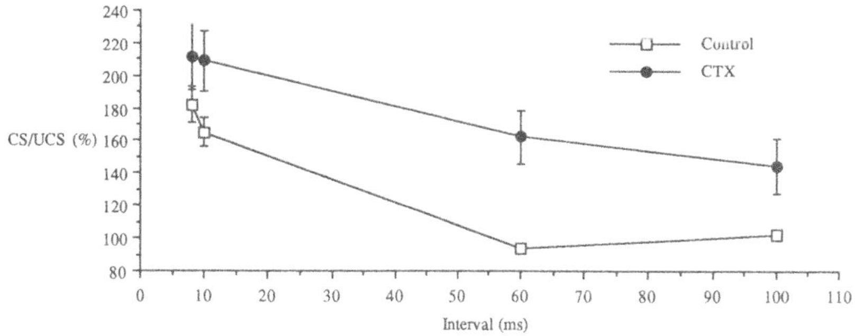

Figure 11.2 Percentage change in amplitude (conditioned stimulus (CS)/unconditioned stimulus (UCS)) of responses to a submaximal stimulus when preceded by a supermaximal conditioning stimulus at interstimulus intervals of 8, 10, 60 and 100 ms. Closed boxes represent the mean of 15 victims of ciguatera poisoning and open circles represent the mean of 15 control subjects. Error bars represent the standard errors of the means. (Reproduced with permission from the *Journal of Neurological Sciences* **101**, 93–97, 1991, Elsevier Science Publishers.)

performed by our group on the effects of ciguatoxin on fish nerves (Capra *et al.*, 1987; Flowers *et al.*, 1987; Flowers, 1989). These studies have shown that fish nerves respond to ciguatoxin in an identical manner to the nerves of rats and humans. Figure 11.3 shows the responses of isolated spinal nerves of a potential carrier fish, Coral Trout, to ciguatoxin. Both the magnitude and duration of the supernormal period are significantly increased in those nerves that were exposed to ciguatoxin.

CONCLUSION

Each year in Australia humans are poisoned after the consumption of toxic fish. The toxin carried by these fish originates in the benthic dinoflagellate *Gambierdiscus toxicus*. Ciguatoxin is a powerful neurotoxin and our group has shown that this toxin alters nerve conduction parameters in the peripheral nerves of rats, humans and fish. The similarities of action in rats and humans indicates that the rat is a useful model for studying human intoxication and may be useful in assessing potential therapeutic agents. The actions of ciguatoxin on fish indicate that fish nerves are susceptible to this toxin and have led us to specu-

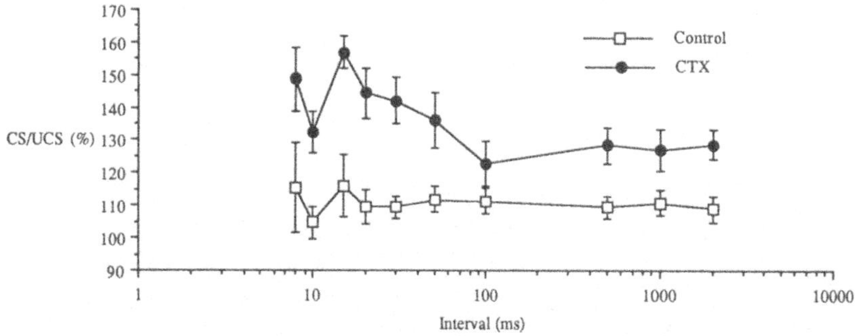

Figure 11.3 Percentage change in amplitude (conditioned stimulus (CS)/unconditioned stimulus (UCS)) of responses to a submaximal stimulus when preceded by a supramaximal conditioning stimulus at a range of intervals from 8–1100 ms. Closed boxes represent the mean of 15 nerves exposed to ciguatoxin and open circles represent the mean of 15 control nerves. Error bars represent the standard errors of the means.

late that some protective partitioning mechanism must exist in fish.

ACKNOWLEDGEMENT

The authors' laboratory has been supported over the past six years by the *Australian Fishing Industry Research and Development Council, the Australian Marine Sciences and Technologies Committee and the Australian Research Council.*

REFERENCES

Anderson, W. (1776) An account of some poisonous fish in the south seas. *Phil. Trans. R. Soc. London* **66**, 544.

Bagnis, R., Kuberski, T. and Laugier, S. (1979) Clinical observations on 3009 cases of ciguatera (fish poisoning) in the South Pacific. *Am. J. Trop. Med. Hyg.* **28**, 1067.

Bagnis, R.A. and Legrand, A-M. (1987) Clinical features of ciguatera (fish poisoning) in French Polynesia. In: *Progress in Venom and Toxin Research*, edited by Gopalakrishnakone, P. and Tan, C.K., p. 372. Singapore: University of Singapore.

Bidard, J.N., Hank, P.M., Vijerberg, C., Chunque, E., Legrand, A-M., Bagnis, R.A. and Sadezeuski, M. (1984) Ciguatoxin is a novel type of sodium channel toxin. *J. Biol. Chem.* **259**, 8353.

Boyarsky, L.L. and Rayner, M.D. (1970) The effects of ciguatoxin on Aplysia neurones. *Proc. Soc. Exp. Biol. Med.* **134**, 332.

Cameron, J. and Capra, M.F. (1990) Neurological studies on the effects of ciguatoxin on mammalian nerves. In: *Ciguatera Seafood Poisoning*, edited by Miller, D. and Tindall, D. Florida: CRC Press (in press).
Cameron, J., Flowers, A.E. and Capra, M.F. (1991a) Effects of ciguatoxin on nerve excitability in rats (Part I). *J. Neurol. Sc.* **101**, 87–92.
Cameron, J., Flowers, A.E. and Capra, M.F. (1991b) Electrophysiological studies on ciguatera poisoning in man (Part II). *J. Neurol. Sc.* **101**, 93–97.
Capra, M.F. (1985) Ciguatera poisoning : Incidence and implications. A final report on Fishing Industry Research Trust Account, Research Grant 82/33. Commonwealth Department of Primary Industry, Canberra.
Capra, M.F. and Cameron, J. (1985) The effects of ciguatoxin on mammalian nerves. *Proc. 5th Int. Coral Reed Cong.* **4**, 457.
Capra, M.F., Flowers, A.E. and Cameron, J. (1987) The effect of ciguatoxin on the rate of Na^+ efflux in unmyelinated olfactory nerves of teleosts. In: *Progress in Venom and Toxin Research*, edited by Gopalakrishnakone, P. and Tan, C.K., p. 418. Singapore: University of Singapore.
Flowers, A.E. (1989) The effects of ciguatoxin on teleost nerves. *MAppSc. Thesis*. Queensland University of Technology, Australia.
Flowers, A.E., Capra, M.F. and Cameron, J. (1987) The effects of ciguatoxin on nerve conduction parameters in teleost fish. In: *Progress in Venom and Toxin Research*, edited by Gopalakrishnakone, P. and Tan, C.K., p. 411. Singapore: University of Singapore.
Gillespie, N.C., Holmes, M.J., Burke, J.B. and Doley, J. (1985) Distribution and periodicity of *Gambierdiscus toxicus* in Queensland, Australia. In: *Toxic Dinoflagellates*, edited by Anderson, D.M., White, A.W., and Baden, D.G., p. 183. New York: Elsevier.
Gillespie, N.C., Lewis, R.J., Pearn, J.H., Bourke, A.T.C., Holmes, M.J., Burke, J.B. and Shields, W.J. (1986) Ciguatera in Australia : Occurrence, clinical features pathophysiology and management. *Med. J. Aust.* **145**, 584.
Kodama, A., Hokama, Y., Yasumoto, T., Fukui, M., Manea, S. and Sutherland, N. (1989) Clinical and laboratory findings implicating palytoxin as a cause of ciguatera poisoning due to *Decapterus macrosoma* (Mackerel) *Toxicon* **27**, 1051.
Lawrence, D.N., Enriquez, M.B., Lumish, R.M. and Maceo, A. (1980) Ciguatera fish poisoning in Miami. *J.A.M.A.* **244**, 254.
Lewis, N.D. (1984) Ciguatera—parameters of tropical health problem. *Human Ecology* **12**, 253.
Lewis, R.J. and Endean, R. (1983) Purification of ciguatoxin-like substances from *Scomberomorus commersoni* and its effect on the rat phrenic nerve diaphragm. *Toxicon Suppl.* **3**, 249.
Maran, J. (1990) Legal implications of ciguatera poisoning. A case history. In: *Ciguatera Seafood Poisoning*, edited by Miller, D. and Tindall, D. Florida: CRC Press (in press).
Miller, D., Dickey, R. and Tindall, D. (1982) The effects of a lipid-extracted toxin from the dinoflagellate *Gambierdiscus toxicus* upon nerve-muscle and intestinal preparations. *Proc. Fed. Am. Soc. Exp. Biol.* **41**, 1561.

Murata, M., Legrand, A-M., Ishibashi, Y. and Yasumoto, T. (1989) Structures of ciguatoxin and its congener. *J. Am. Chem. Soc.* **111**, 8929.
Ragelis, E.P. (1984) Ciguatera Seafood Poisoning, Overview. In: *Seafood Toxins*, edited by Ragalis, E. A., p. 25. Washington DC: Chem. Soc. Symp. Series 262.
Tachibana, K. (1980) Structural studies on marine toxins. *PhD Thesis*, Hawaii: University of Hawaii.
Taylor, F.J.R. (1979) A description of the benthic dinoflagellate associated with maitotoxin and ciguatoxin, including observations on Hawaiian material. In: *Toxic Dinoflagellate Blooms*, edited by Taylor, D.L. and Seliger, H.S., p.71. New York: Elsevier Nth Holland.
Taylor, F.J.R. (1984) Human and domestic animal fatalities as well as skin reactions associated with the rough trigger fish (*Canthidermis maculata*) in Dominica, West Indies following hurricane David. *Rev. Int. Oceanogr. Med.* 73–74.
Tonge, J.I., Battey, Y., Forbes, J.J. and Grant, E.M. (1967) Ciguatera poisoning: A report of two outbreaks and a probable fatal case in Queensland. *Med. J. Aust.* **2**, 1088.
Withers, N.W. (1982) Ciguatera fish poisoning. *Am. Rev. Med.* **33**, 97.
Yasumoto, T., Bagnis, R.A., Thevenin, S. and Garcon, M. (1977a) A survey of comparative toxicity in the food chain of ciguatera. *Bull. Jpn. Soc. Sci. Fish.* **43**, 1015.
Yasumoto, T., Nakajima, I., Bagnis, R.A. and Adachi, R. (1977b) Finding a dinoflagellate as a likely culprit of ciguatera. *Bull. Jpn. Soc. Sci. Fish.* **43**, 1021.
Yasumoto, T., Raj, V. and Bagnis, R. (1984) *Seafood poisoning in Tropical Regions*. Faculty of Agriculture Tohubu University, Miyagi.

12 FUNNEL-WEB SPIDER TOXINS

GRAHAM M. NICHOLSON, IAN SPENCE, ALFIO COMIS, MARGARET I. TYLER AND MERLIN E. H. HOWDEN

INTRODUCTION

Funnel-web spiders are a group of aggressive, nocturnal and particularly venomous arachnids restricted to south-eastern Australia. They occur from eastern Tasmania north along the eastern coast to south-east Queensland and an isolated area on the Eyre peninsula in South Australia. There are two medically important genera of funnel-web spiders. Recently a taxonomic revision of the 35 funnel-web spider species has left only three remaining in the genus *Atrax*, which was the genera used for all funnel-web spiders in earlier years, whilst the others have been placed into six species groups within the genus *Hadronyche*.

Spider venoms are known to contain several distinct classes of toxins. The funnel-web spider venom appears, however, to be unique, producing marked changes in the activity of peripheral nerves by mechanisms unlike those of other spider neurotoxins. Two species have been studied for their neurotoxic actions, the notorious Sydney funnel-web spider *Atrax robustus*, which is limited to a radius of 160 km from the center of the city of Sydney, and the Blue Mountains Funnel-web *Hadronyche versutus* which is confined to the Blue Mountains west of Sydney. *Atrax robustus* (Figure 12.1) is remarkable in that of the 14 fatalities so far reported since 1927 all have been due to the male *A. robustus* rather than the female which is the case with all other highly poisonous spiders. In fact the male venom is approximately six

Figure 12.1 Sydney funnel-web spider (*Atrax robustus*).

times more toxic than female venom. Fortunately the male is easily recognized by the presence of pointed tibial spurs on the second pairs of legs.

The original work was carried out in 1934 by Kellaway, who showed that mice and guinea pigs bitten by *A. robustus* failed to die. More recently it has been proven that it adversely affects only primates such as man and monkeys. Fortunately an effective antivenom has been available since late 1980.

NEUROTOXIC ACTIONS OF THE VENOM

The most obvious action of funnel-web spider venom is the appearance of generalized muscle twitches. The available evidence suggests that these fasciculations appear to be due to activity in the motor nerves rather than a direct action on skeletal muscle. At a concentration 10 mg/ml the venom of *A. robustus* was shown to cause repetitive firing in nerves and to produce spontaneous end-plate potentials in skeletal muscle maintained *in vitro* (Spence, 1979). The fasciculations of skeletal muscle fibres produced by the venom can also be blocked when the muscle is bathed in (+)-tubocurarine. These effects all indicate that the venom has an axonal site of action and that the fasciculations probably arise from the spontaneous release of acetylcholine from the motor nerve terminal rather than a change in the contractile properties of the muscle.

NEUROTOXIC ACTIONS OF FUNNEL-WEB SPIDER TOXINS

Most of our current work has centered on the use of two toxins, robustoxin (RTX) and versutoxin (VTX), isolated from *A. robustus* and *H. versutus* respectively. Both toxins have had their amino acid sequence fully identified (Sheumack *et al.*, 1985; Brown *et al.*, 1988). They are both 42 amino acid polypeptides with similar sequences (eight substitutions) containing four disulphide bridges (see Table 12.1). This indicates that the toxin is very tightly bound together. Interestingly the venom of the female *A. robustus* was found to lack RTX and may account for its apparent lack of lethality. More recently RTX has been synthesized and refolded into its native conformation.

To assess if the purified and synthetic toxins possessed the toxic action of the native venom the actions of the toxins on axonal excitability were tested *in vitro*. Similar to the whole venom both RTX and VTX produced an increase in the amplitude and duration of the compound action potential in both sensory and motor nerves at concentrations of 1 mg/ml. The synthetic refolded RTX also produced similar actions while the straight chain synthetic RTX lacked any observable activity but interestingly blocked the action of RTX.

To investigate if the repetitive firing caused by the venom is confined only to axonal membranes, the direct action of the toxins on skeletal muscle was assessed in a Cl^- reduced Krebs

Table 12.1 Amino acid sequence of versutoxin and robustoxin

	10 20 30 40
Robustoxin	CAKKRNWCGK**N**EDCCCPMKC**I**YAWYN**Q**QGSC**ETTITG**L**F**KKC
Versutoxin	CAKKRNWCGK**T**EDCCCPMKC**V**YAWYN**E**QGSC**QSTISA**L**W**KKC

solution (by substitution with SO_4^{2-}). This renders skeletal muscle more "excitable" (Vyskocil, 1977). Indirect actions of the toxin were blocked using (+)-tubocurarine. RTX produced a lengthening of the relaxation phase of muscle contraction to direct muscle stimulation indicating the action of the toxin is not confined to neuronal membranes. In further experiments using standard intracellular recording techniques this prolongation of the relaxation phase was found to be due to repetitive firing of action potentials from a single stimulus rather than a prolongation of the muscle action potential (see Figure 12.2).

These effects are consistent with the belief that funnel-web spider venom increases the probability of sodium channels opening possibly by altering the electric field within the membrane (Spence, 1979). The effect of this is to produce spontaneous action potentials in nerve fibres which leads to muscle fasciculation.

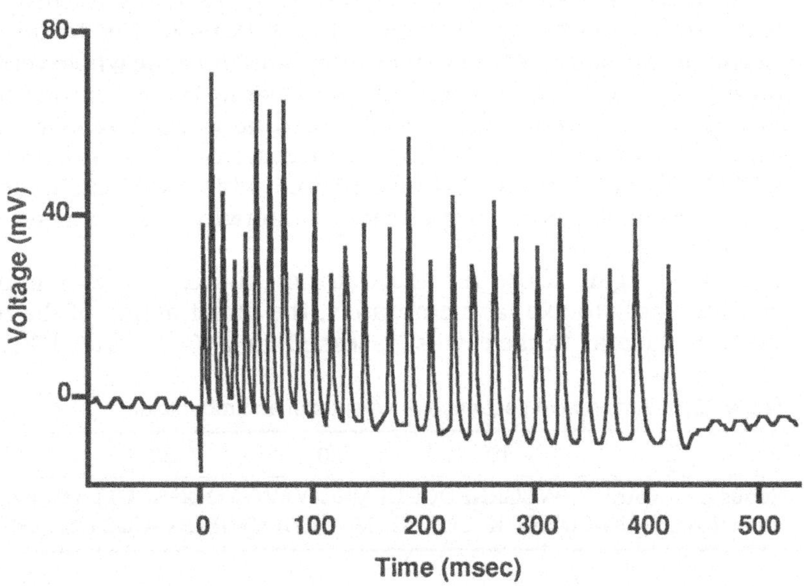

Figure 12.2 Effect of robustoxin (10 μg/ml) on rat diaphragm muscle.

REFERENCES

Brown, M.R., Sheumack, D.D., Tyler, M.I. and Howden, M.E.H. (1988) Amino acid sequence of versutoxin, a lethal neurotoxin from the venom of the funnel-web spider *Atrax versutus*. *Biochem. J.* **250**, 401–405.

Kellaway, C.H. (1934) A note on the venom of the Sydney funnel-web spider *Atrax robustus*. *Med. J. Aust.* **1**, 678–679.

Sheumack, D.D., Claassens, R., Whiteley, N.M. and Howden, M.E.H. (1985) Complete amino acid sequence of a new type of lethal neurotoxin from the venom of the funnel-web spider *Atrax robustus*. *FEBS Lett.* **181**, 154–156.

Spence, I. (1979) Electrophysiological studies of the venoms of the Funnel-web spider *Atrax robustus* and the tick *Ixodes holocyclus*. In: *Neurotoxins: Fundamental and Clinical Advances*, edited by Chubb, I.W. and Geffen, L.B. pp. 161–175. Adelaide: Adelaide Union University Press.

Sutherland, S.K. (1983) Genus *Atrax* Cambridge, the funnel-web spiders. In: *Australian Animal Toxins*, pp. 257–298. Melbourne: Oxford University Press.

Vyskocil, F. (1977) Diazepam blockade of repetitive action potentials in skeletal muscle fibres. A model of its membrane action. *Brain Res.* **133**, 3315–3328.

13 TETANUS TOXIN
Its Effect on Synaptosomal Protein Phosphorylation

PETER R. DUNKLEY, PETER P. PRESEK, FLORIAN DREYER AND PAULA E. JARVIE

INTRODUCTION

Tetanus toxin is produced by *Clostridium tetani* and its clinical effects are due to inhibition of neurotransmitter release at both central inhibitory synapses and at the neuromuscular junction (Habermann and Dreyer, 1986). The toxin is a protein, consisting of a heavy and a light chain linked by disulphide bonds (Weller *et al.*, 1989), and is a member of the family of clostridial neurotoxins, including the botulinum toxins, which share structural and functional similarities. It is known that these toxins bind to neuronal cell membranes, become internalized, may be processed for example by proteolysis or reduction, and act irreversibly to inhibit the release of a number of neurotransmitters (Habermann and Dreyer, 1986). The exact site at which tetanus toxin acts is unknown, but it does not appear to modulate calcium entry into nerve terminals (Penner *et al.*, 1986 and Gansel *et al.*, 1987).

When neurons are depolarized a number of molecular events are observed which are dependent on calcium entry and which occur concomitantly with neurotransmitter release. Thus, an increase or decrease in the extent of phosphorylation of a number of nerve terminal proteins is seen in isolated synaptosomes (Dunkley and Robinson, 1986). It was found that botulinum toxin A interferes with the phosphorylation of proteins in pure cholinergic synaptosomes obtained from the electric organ of *Torpedo* (Guitart *et al.*, 1987), but the function of the phosphoproteins affected was not known. The major phosphoprotein seen in synaptosomes from rat central nervous system is synapsin I (Dunkley and Robinson, 1986). This protein binds to the surface of synaptic vesicles and to the actin cytoskeletal network (De Camilli and Greengard, 1986). Increases in phosphorylation of synapsin I cause it to dissociate from vesicle membranes, which allows the vesicles to dock with the plasma membrane and release their neurotransmitters (Llinas *et al.*, 1985; Bahler and Greengard, 1987).

As a working hypothesis we propose that tetanus toxin acts to decrease the phosphorylation of synapsin I in nerve terminals leading to an inhibition of neurotransmitter release. This study was aimed at testing the first part of this hypothesis.

METHODS

Synaptosomes isolated from rat cerebral cortex on Percoll density gradients were washed twice in Krebs' buffer containing 0.1 mM calcium (Dunkley *et al.*, 1988). Tetanus toxin was a gift from Professor Habermann of Giessen University and was freshly dialyzed and then diluted prior to use with phosphate-buffered-saline containing 0.1% BSA. Inactivation of the toxin, when necessary, was achieved by heating in a boiling water bath for 15 min. Toxin was added to synaptosomes (5 mg/ml protein) at various concentrations (0.01–10 μg/ml) for various times (0–60 min) at 37°C. After washing the synaptosomes protein phosphorylation was initiated by the addition of ^{32}Pi (finally 0.75 mCi/ml) for 45 min. Aliquots were then incubated with an equal volume of either control or depolarizing (41 mM K^+) buffer for 15 s after which proteins were fractionated by PAGE and phosphoproteins were detected by autoradiography (Dunkley and Robinson, 1986). Results for synapsin I phosphorylation were expressed as percentages relative to non-depolarized samples which had not been exposed to toxin. Synapsin I is a doublet with apparent molecular weights of 86 and 75 kD (Dunkley and Robinson, 1986) and only the 75 kD band has been quantitated for these studies.

RESULTS

Depolarization of synaptosomes led to an increase in the extent of phosphorylation of synapsin I and a decrease in P96 labelling (Figure 13.1, tracks 1,2). Incubating synaptosomes for 60 min with increasing concentrations of tetanus toxin led to a pro-

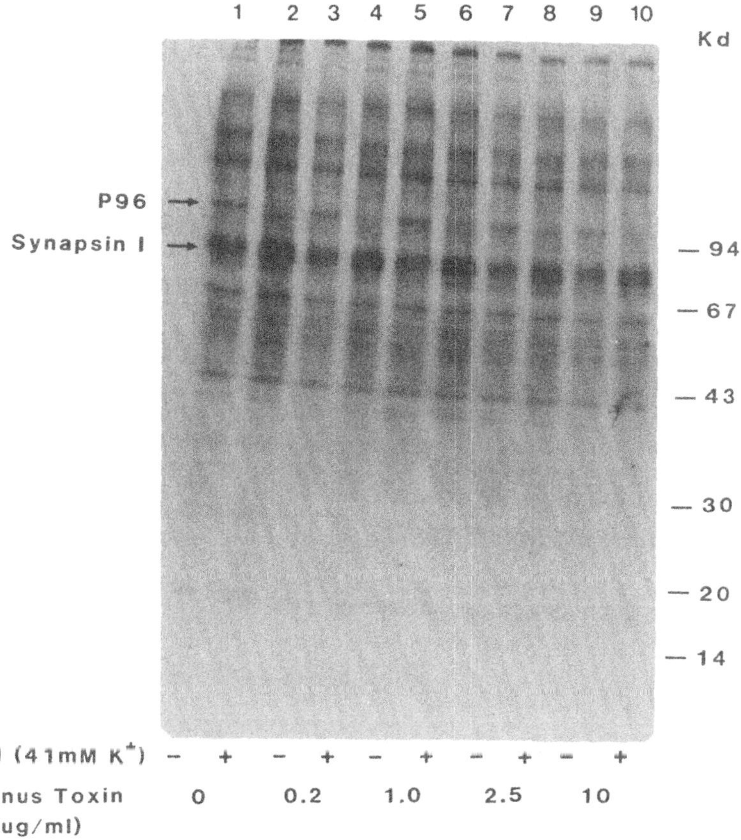

Figure 13.1 The effect of tetanus toxin on synaptosomal protein phosphorylation. Synaptosomes isolated from rat cerebral cortex were incubated with concentrations of tetanus toxin from 0–10 µg/ml for 60 min at 37°C. ^{32}Pi was added for 45 min before synaptosomes were exposed to control (–) or depolarizing high K^+ buffer (+) for 15 s at 37°C. After fractionation of proteins by SDS–PAGE, autoradiographs were prepared. The location of P96 and synapsin I (LHS) and the position of the molecular weight markers (RHS) are shown.

gressive decrease in synapsin I labelling in the depolarized samples (Figure 13.1, tracks 4,6,8,10). This effect was first observed at 0.2 µg/ml, but the maximum decrease of 54% was seen at 10 µg/ml. The decrease in P96 phosphorylation on depolarization was not affected by the presence of the toxin. No significant effect of tetanus toxin was observed in the non-depolarized samples (Figure 13.1).

To ensure that these effects of tetanus toxin were compatible with its normal biological action of inhibiting release and were not due to some non-specific toxicity two experiments were undertaken. First, heat inactivated toxin was compared with native toxin and no decrease in synapsin I phosphorylation occurred (not shown). Second, it was found that incubating synaptosomes with tetanus toxin at 10 μg/ml for up to 30 min had no effect on synapsin I phosphorylation in either control or depolarized conditions but longer times led to a decreased level of labelling only in the depolarized samples (Figure 13.2).

When lyzed synaptosomes were incubated with [C-^{32}P] ATP under optimum conditions for cyclic AMP-stimulated protein kinase, protein kinase C or calmodulin kinase II, tetanus toxin at 10 μg/ml was found to have no effect on endogenous protein phosphorylation (not shown).

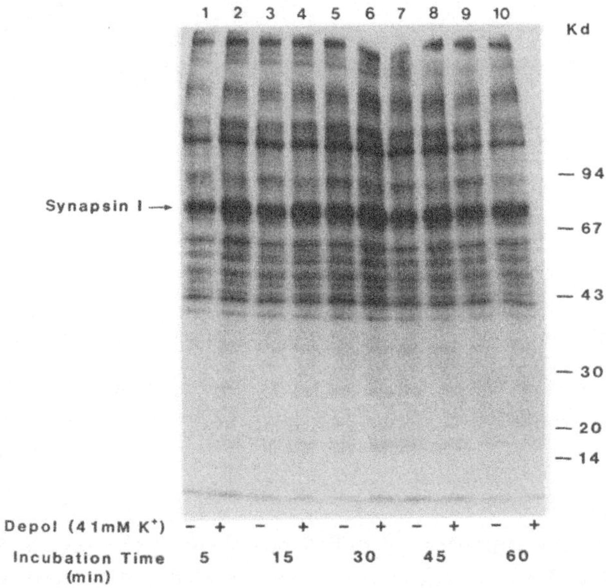

Figure 13.2 The effect of incubation time on tetanus toxin action. The experiment shown in Figure 13.1 was repeated for toxin at 10 μg/ml which had been incubated with synaptosomes for the times indicated, prior to labelling with ^{32}Pi.

CONCLUSIONS

The change in phosphorylation of synapsin I normally observed on depolarization of synaptosomes was decreased by tetanus toxin. Such an effect on protein phosphorylation could have been caused by non-specific damage to synaptosomes, a decrease in the activity of one or more protein kinases, an increase in protein phosphatase activity, modification of ATP or the phosphoprotein substrate levels, or by modification of the level of kinase effectors such as calcium, cyclic nucleotides or inositol phospholipids.

A non-specific toxicity seems unlikely as basal levels of protein phosphorylation were unaffected by tetanus toxin and heat inactivated toxin had no effect on depolarized synaptosomes. Furthermore, inhibition of stimulated synapsin I phosphorylation did not occur in the first 30 min of incubation. These findings are similar to those seen with tetanus toxin action on release of neurotransmitters (Habermann and Dreyer, 1986) and are consistent with the requirement for tetanus toxin internalization prior to its action.

Tetanus toxin did not directly alter protein kinase activity. However, it may be that a processing step is required for the action of the toxin, which normally occurs within nerve terminals, and which does not occur in the lyzed synaptosomes under the conditions required for protein phosphorylation. Thus fragments of the toxin, such as the heavy and light chains, should be tested for their effects on protein kinase activity (Ahnert-Hilger *et al.*, 1989; Weller *et al.*, 1989).

It seems unlikely that the depolarization-stimulated uptake of calcium into synaptosomes is affected by tetanus toxin because the dephosphorylation of P96 still occurred in the presence of high levels of toxin. This dephosphorylation is completely dependent on extracellular calcium (Robinson *et al.*, 1987).

Further work is required to identify the site(s) at which tetanus toxin is acting to decrease synapsin I phosphorylation. Whatever the mechanism, its elucidation will provide insight into the relationship between protein phosphorylation, neurotransmitter release and the molecular mode of action of tetanus toxin.

ACKNOWLEDGEMENTS

The National Health and Medical Research Council and the Newcastle University Research Committee are thanked for supporting this project. Dr D. Findik is thanked for reviewing the manuscript and Mrs Ena Mawer for typing it.

REFERENCES

Ahnert-Hilger, G., Weller, U., Dauzenroth, M.E., Habermann, E. and Gratzel, M. (1989) The tetanus toxin light chain inhibits exocytosis. *FEBS Letters* **242**, 245–248.

Bahler, M. and Greengard, P. (1987) Synapsin I bundles F-actin in a phosphorylation-dependent manner. *Nature* **326**, 704–707.

Browning, M.D., Huganir, R. and Greengard, P. (1985) Protein phosphorylation and neuronal function. *J. Neurochem.* **45**, 11–23.

Burgoyne, R.D. (1984) Mechanisms of secretion from adrenal chromaffin cells. *Biochem. Biophys. Acta* **779**, 201–206.

De Camilli, P. and Greengard, P. (1986) Synapsin 1: A synaptic vesicle-associated neuronal phosphoprotein. *Biochem. Pharmac.* **35**, 4349–4357.

Dunkley, P.R., Heath, J.W., Harrison, S.M., Jarvie, P.E., Glenfield, P.J. and Rostas, J.A.P. (1988) A rapid Percoll gradient procedure for isolation of synaptosomes directly from an S_1 fraction: homogeneity and morphology of subcellular fractions. *Brain Research* **441**, 59–71.

Dunkley, P.R. and Robinson, P.J. (1986) Depolarisation-dependent protein phosphorylation in synaptosomes: mechanisms and significance. *Progress in Brain Research* **69**, 273–293.

Gansel, M., Penner, R. and Dreyer, F. (1987) Distinct sites of action of clostridial neurotoxins revealed by double-poisoning of mouse motor nerve terminals. *Pflugers Arch* **409**, 533–539.

Guitart, X., Egea, G., Salsona, C. and Marsal, J. (1987) Botulinum neurotoxin inhibits depolarisation-stimulated protein phosphorylation in pure cholinergic synaptosomes. *FEBS Letters* **219**, 219–223.

Habermann, E. and Dreyer, F. (1986) Clostridial neurotoxins: handling and action at the cellular and molecular level. *Current Topics in Microbiology and Immunology* **129**, 93–179.

Llinas, R., McGuinness, T., Leonard, C.S., Sugimari, M. and Greengard, P. (1985) Intraterminal injection of synapsin I or calcium/calmodulin-dependent protein kinase II alters neurotransmitter release at the squid giant synapse. *Proc. Natl. Acad. Sci.* **82**, 3035–3039.

Penner, R., Neher, E. and Dreyer, F. (1986) Intracellularly injected tetanus toxin inhibits exocytosis in bovine adrenal chromaffin cells. *Nature* **324**, 76–78.

Robinson, P.J., Hauptschein, R., Lovenberg, W. and Dunkley, P.R. (1987) Dephosphorylation of synaptosomal proteins p96 and p136 is regulated by both depolarisation and calcium, but not by a rise in cytosolic calcium alone. *J. Neurochem.* **48**, 187–195.

Weller, U., Dauzenroth, M.E., Meyer zu Heringsdorf, D. and Habermann, E. (1989) Chains and fragments of tetanus toxin. Separation, reassociation and pharmacological properties. *Eur. J. Biochem.* **182**, 649–656.

14 THE EXOTOXIN OF MELIOIDOSIS, A DISEASE OF ANIMALS AND MAN

RAHMAH MOHAMED, MAHESWARAN PERIASAMY, GHAZALLY ISMAIL AND NOOR EMBI

INTRODUCTION

Pseudomonas pseudomallei, a Gram-negative soil bacterium found in Asia and tropical Australia is the causative organism of melioidosis. Clinical and pathological manifestations of melioidosis in man and animals are varied, the subclinical form of melioidosis is most common in animals whilst the acute septicaemic form of the disease resulted in high mortality in humans (Smith *et al.*, 1987). Culture filtrates of *P. pseudomallei* contain a number of extracellular substances which exhibited a variety of biological effects such as lethality, necrotizing, proteolytic and lecithinase activity (Nigg *et al.*, 1955). Accumulating evidence suggests that the trypsin-sensitive, heat-labile lethal exotoxin plays an important role in melioidosis (Wasiman *et al.*, 1987 and Mohamed *et al.*, 1988a). Mice immunized with the heat-denatured exotoxin of *P. pseudomallei* were protected from the lethal effects of infection of *P. pseudomallei* (Smith *et al.*, 1987). The demonstration of serum exotoxin antibody in animals naturally exposed to *P. pseudomallei* infection strongly implicates the importance of exotoxin in the pathogenesis of melioidosis (Ghazally *et al.*, 1989).

Previous studies in our laboratory have addressed the possible molecular mechanism by which *P. pseudomallei* exotoxin functions. Exotoxin at concentrations > 20 mg/ml resulted in complete fragmentation of cells in culture (Maheswaran *et al.*, 1987).

The cytotoxic activity of the exotoxin as evaluated by morphological examinations and inhibition of DNA synthesis was observed in a wide range of mammalian cell lines (Mohamed *et al.*, 1988a). Recent studies demonstrated that protein synthesis in macrophages was inhibited with as low as 0.05 mg/ml of the lethal exotoxin (Mohamed *et al.*, 1988b). Under these conditions cells were viable and capable of maximal synthesis of DNA. Nevertheless, the mechanism by which the lethal exotoxin inhibits cell protein synthesis has yet to be elucidated.

The intracellular activities of many bacterial toxins are enzymatic and in many instances involve similar biochemical events. This report describes the NAD^+-dependent inhibitory effect of *P. pseudomallei* exotoxin on protein synthesis in rabbit reticulocyte lysate and the involvement of ADP-ribosylation in the cellular mechanism of *P. pseudomallei* exotoxin.

MATERIALS AND METHODS

Isolation of *P. pseudomallei* Exotoxin

P. pseudomallei was grown in Brain Heart Infusion Broth at 32°C for seven days and the crude exotoxin was isolated from the culture filtrate as described previously (Wasiman *et al.*, 1987). The exotoxin preparation was subjected to further purification by ion-exchange chromatography (CM-Sephadex) in 50 mM phosphate buffer, pH 6.4. Pure exotoxin was eluted with the above buffer containing 0.5M NaCl.

Limited Digestion of the Exotoxin

Limited digestion of the purified exotoxin (200 mg/ml) by endogeneous protease(s) was performed by incubating the exotoxin at 37°C at varying times in phosphate saline buffer pH 7.2.

Limited digestion of the exotoxin by trypsin was performed by incubating the toxin (300 mg/ml) with 0.4 mg/ml of trypsin at 25°C for varying times in the above buffer. Proteolysis was terminated by the addition of soya bean trypsin inhibitor (0.6 mg/ml).

Preparation of Rabbit Reticulocyte Lysate

Reticulosis was induced in rabbits by a modification of the method described by Clemens (1984). Reticulocytes from acetalphenylhydrazine-treated rabbits were washed twice with 0.14 M

NaCl containing 5 mM KCl and 1.5 mM $MgCl_2$. Subsequently 0.6 vol of 5 M sucrose containing 0.15M KCl was added to stop lysis and the suspension was allowed to stand for 2 min. After centrifugation at 15 000g for 15 min to remove unlysed cells and cell membrane fragments, the supernatant was collected and stored (1.0 ml aliquots) in liquid nitrogen. Reticulocyte lysates depleted of NAD^+ were prepared by desalting the lysate through a Sephadex G-25 column in 5 mM Hepes buffer, pH 7.2 containing 0.14 M NaCl, 5 mM KCl, 5 mM magnesium acetate and 5 mM glucose.

In vitro Protein Synthesis

Protein synthesis in lysate was assayed by measuring the incorporation of [^{14}C] leucine (Clemens, 1984). In experiments to investigate the effect of the exotoxin on protein synthesis, varying doses of the exotoxin were employed. The reaction mixture (100 ml) contained 75 mM KCl, 2 mM magnesium acetate, 3 mM glucose, 1 mM ATP, 0.2 mM GTP, 10 mM Tris-HCl (pH 7.6), 7 mM creatine phosphate, 1 mg/ml creatine phosphokinase, 20 mM haemin and 50–200 mM amino acids mixture containing 5 mCi/ml [^{14}C] leucine (sp. activity 50 mCi/mmol). After incubation at 37°C for 60 min, 50 mg of BSA was added and the reaction was stopped by the addition of 20% TCA. The mixture was heated to 90°C for 15 min to hydrolyze charged t-RNA before being filtered on a GF/C filter paper.

Radioactivity was determined using the scintillation counter. Protein synthesis was expressed as a percentage of the value obtained for the reaction in the absence of the exotoxin.

Protein Synthesis in HeLa Cells

HeLa cells, strain S-3, were grown in 24-well dishes at 2×10^5 cells/ml. After incubation with exotoxin for 1 h, cells were pulsed for 1 h with 5 mCi/ml ^{14}C leucine (sp. activity 50 mCi/mmole). The radioactivity of acid-insoluble materials was measured as described above.

ADP-Ribosylation Activity

Transferase II was purified from rabbit reticulocyte lysate as described elsewhere. The ammonium sulphate (40%–60%) precipitate of the lysate obtained upon centrifugation at 100 000 g (SW27) was chromatographed on Sephadex G100. Fractions containing GTP-binding activity were pooled and exhaustively dialyzed in 20 mM Tris-HCl pH 7.5 containing 0.1 mm EDTA, 1 mm DTT buffer. Polyacrylamide gel electrophoresis in the presence of SDS showed that the purified Transferase II was of molecular weight 14 kD.

Nicotinamide adenine dinucleotide (NAD) transferase activity was assayed in a mixture containing 50 mM Tris-HCl (pH 8.2), 0.1 mM EDTA, 40 mM DTT, 12.5 pmoles Transferase II, nicotinamide [U-^{14}C] adenine dinucleotide ([^{14}C]NAD^{+} 41 mCi/mmole). After incubation, 10% TCA was added in the presence of BSA, acid precipitable material obtained was washed and radioactivity was determined in the scintillation counter.

RESULTS

Inhibition of Protein Synthesis by the Intact Exotoxin

The effect of *Pseudomonas pseudomallei* exotoxin on cell protein synthesis was studied using a number of mammalian cell lines (Figure 14.1). Inhibition of protein synthesis upon exposure to the intact exotoxin is in a dose dependent manner with maximum inhibition (80%) at 0.5 mg/ml exotoxin. Under these conditions cells were viable and capable of maximal synthesis of DNA. This suggests that the intoxication of HeLa, Vero and Hep-2 upon exposure to the exotoxin is most likely attributed to the inhibition of protein synthesis. Figure 14.2 compares the profiles of protein synthesis in the lysate and HeLa cells incubated with varying doses of the intact exotoxin. Although inhibition of protein synthesis in HeLa cells is in a dose-dependent manner, inhibition of protein synthesis was not apparent in reticulocyte lysate upon exposure to 0.1–0.5 mg/ml of the intact exotoxin. Slight inhibition of protein synthesis (25%) was observed at the higher concentration of the exotoxin (2.0 mg/ml).

Inhibition of Protein Synthesis in Lysate

The inhibitory effect of the exotoxin on cell-free protein synthesis was only apparent when exotoxin was initially subjected to limited digestion with trypsin or endogenous protease. Exotoxin preincubated with trypsin for 60 min caused (70%) inhibition of protein synthesis (Figure 14.3a). It may be inferred that the observed inhibition of protein synthesis in the lysate is attributed to cleavage products of the exotoxin. Exotoxin preparations pre-

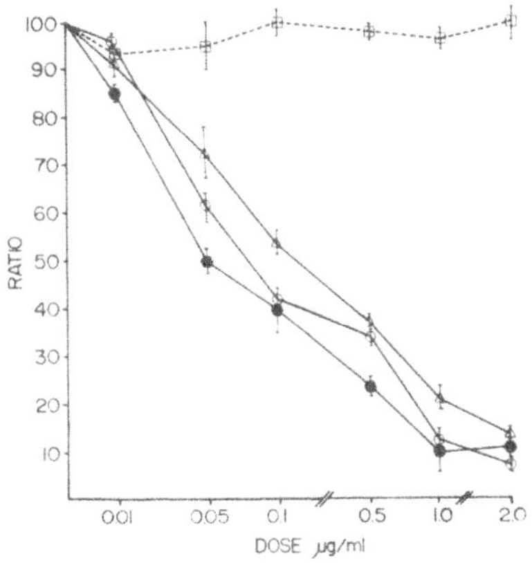

Figure 14.1 Protein synthesis in HeLa (●), Vero (o) and Hep-2 cells (Δ) after incubation with the exotoxin and heat denatured exotoxin (-□-). Ratio refers to incorporation of radioactivity into treated cells divided by incorporation into untreated cells. Each point represents the mean value from three experiments carried out in duplicate. Error bars are standard error of the mean (S.E.M.).

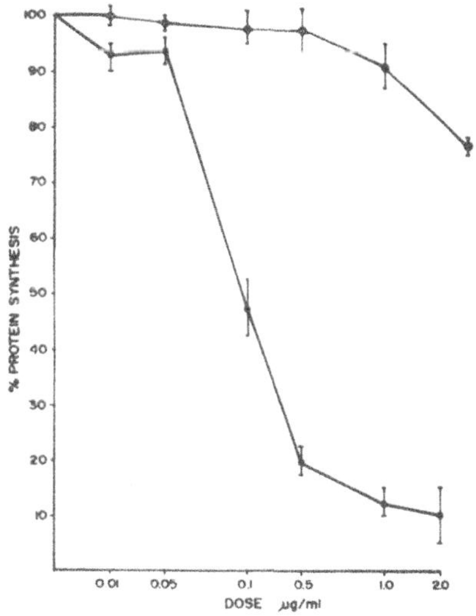

Figure 14.2 Effect of varying doses of intact exotoxin of *P. pseudomallei* on [^{14}C] leucine incorporation into proteins in HeLa cells (●) and rabbit reticulocyte lysate (o). % protein synthesis refer to ratio described above. Error bars are stand error of the mean (S.E.M.).

incubated at 37°C for 6 h also resulted in inhibition of cell-free protein synthesis (Figure 14.3b). Under this condition, cleavage products are presumably released by the endogenous proteolytic activity contaminating the exotoxin preparation.

Requirement for NAD$^+$

Inhibition of protein synthesis in lysate by the exotoxin is NAD$^+$-dependent. Figure 14.4 shows that inhibition of protein synthesis by the nicked exotoxin occurs only in the presence of NAD$^+$, whereas no inhibition was seen in lysate depleted of NAD$^+$ (Figure 14.5). The slight inhibition observed in the lysate not depleted of NAD$^+$ is attributed to the presence of endogenous NAD$^+$. Results

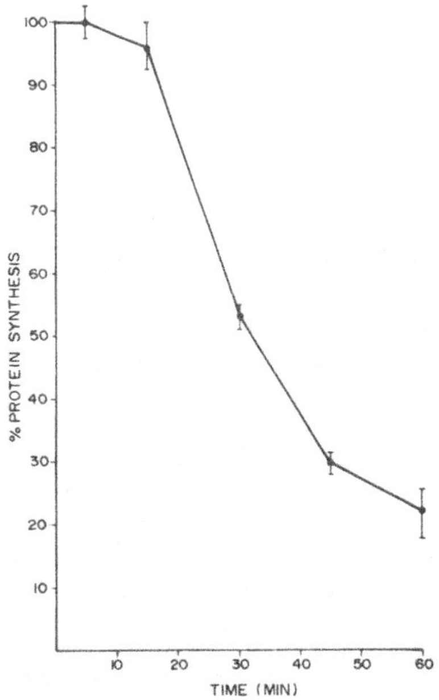

Figure 14.3a The effect of endogeneous protease activity on protein synthesis of rabbit reticulocyte lysate. Intact exotoxin was incubated with trypsin (0.4 µg/ml) as described.

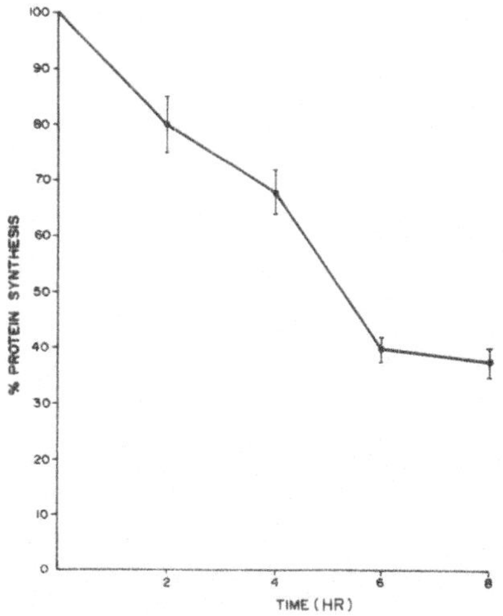

Figure 14.3b The effect of endogeneous protease activity on protein synthesis of rabbit reticulocyte lysate. Intact exotoxin was incubated at 37°C.

obtained show that inhibition of protein synthesis occurred in the presence of 0.01–1 mM NAD^+.

ADP-Ribosylating Activity of the Exotoxin

Figure 14.6, shows that nicked exotoxin is enzymatically active in transferring radioactivity from [^{14}C] NAD^+ into acid-precipitable proteins in rabbit reticulocyte lysate. However no incorporation of [^{14}C] was observed when [^{14}C] NAD^+ labelled at the nicotinamide was employed, showing that the ADP-ribose moiety of NAD^+ was transferred. This suggests that the exotoxin has an ADP-ribosylating activity. ADP-ribosylating activity was evident as early as 1 min upon incubation with [^{14}C] NAD^+ and plateaus after 2 min. A time-course profile of [^{14}C] incorporation in the lysate showed no increase in ^{14}C incorporation at incubation times up to 15 min.

ADP-Ribosylation of Transferase II

The ADP-ribosylating activity of *P. pseudomallei* exotoxin was also seen when transferase II purified from rabbit reticulocyte lysate was the substrate. Incorporation of ^{14}C from [^{14}C] NAD^+, uniformly labelled in the adenosine, into acid-insoluble material occurred as early as 0.5 min upon incubation with the nicked exotoxin (2 mg/ml) (Figure 14.7). As above, evidence that the *P. pseudomallei* exotoxin transferred the ADP-ribose moiety of NAD^+ to Transferase II, was shown by an experiment using [carbonyl-^{14}C] nicotinamide adenosine dinucleotide substituted for [^{14}C] NAD^+ in the ADP-ribosylation assay. When [carbonyl-^{14}C] NAD^+ was employed, no label was transferred to the acid-precipitable material (Table 14.1). The data suggest that the *P. pseudomallei* exotoxin catalyzed ADP-ribosylation of NAD^+ of Transferase II.

DISCUSSION

Although melioidosis and its causative organism *P. pseudomallei* was first recognized 75 years ago by Whitmore and Krishnaswani (1912), little is known about its pathogenic mechanism of infection. The role of the exotoxin in melioidosis has not been directly established, however its involvement has been suggested in experiments using laboratory animals (Nyonya *et al.*, 1986). Recently, data seem to suggest an *in vivo* pathogenic role of the exotoxin during the natural infection as well as the demonstrations that the exotoxin is immunogenic *in vivo* (Ghazally *et al.*, 1989). Using a competitive ELISA to determine levels of exotoxin in serum of sheep, we showed that in sheep diagnosed as death due to me-

Table 14.1 ADP-ribosylation of Transferase II by nicked exotoxin of *P. pseudomallei*

Exotoxin added (2μg/ml)	TCA precipitate (CPM) [^{14}C]NAD^+ (Label at adenosine)	TCA precipitate (CPM) [^{14}C]NAD^+ (Label at nicotinamide)
a. Nicked exotoxin	3 901	558
b. Denatured exotoxin	453	433
c. None	253	324

Exotoxin was incubated in reactions containing Transferase II and incubated for 15 min at 23°C to obtain maximum transfer of ADP-ribose to Transferase II.

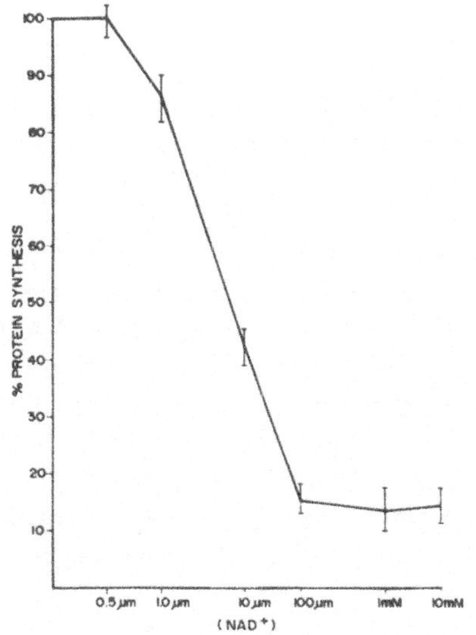

Figure 14.4 The effect of nicked exotoxin (2μg/ml) and varying concentration of NAD^+ on protein synthesis in a reticulocyte lysate.

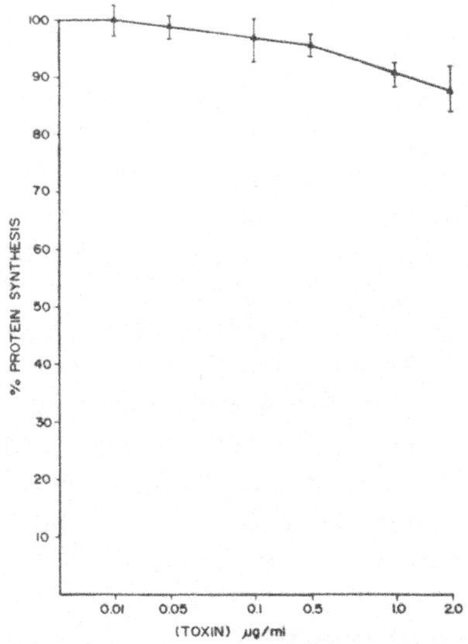

Figure 14.5 Dose response profile of protein synthesis in lysate depleted of NAD^+.

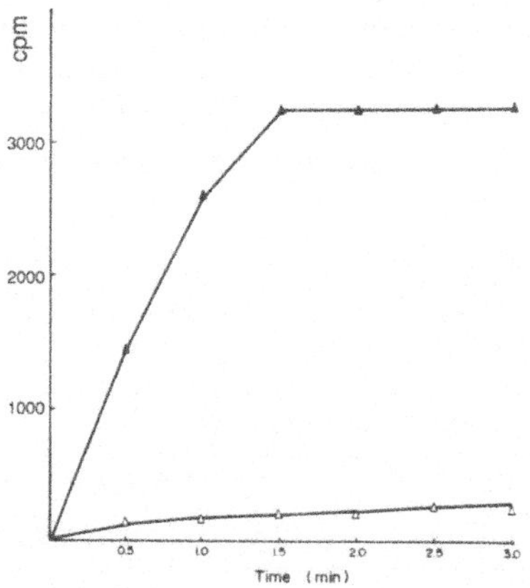

Figure 14.6 ADP-ribosylating activity of the nicked exotoxin in rabbit reticulocyte lysate. CPM indicate incorporation of $[^{14}C]NAD^+$ into acid precipitable material. Toxin (black triangle), –Toxin (white triangle).

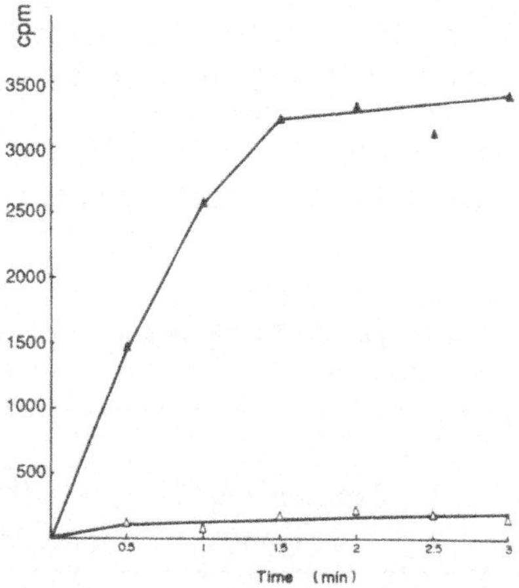

Figure 14.7 ADP-ribosylating activity of the nicked exotoxin using Transferase II as substrate. Toxin (black triangle), –Toxin (white triangle).

lioidosis, the amount of exotoxin in serum is around 40–50 mg/ml serum (unpublished data). This observation perhaps reiterates the observation of toxaemia in animals infected with *P. pseudomallei* reported by Danneberg and Scott (1958) 27 years ago. The present study provides further evidence on the role of the exotoxin as an important virulence determinant in the pathogenecity of *P. pseudomallei*.

ADP-ribosylation, the transfer of the ADP-ribose moiety of NAD^+ to a macromolecule was first reported by Hayaishi *et al.* in 1968 as the mechanism of the intoxication of diphtheria toxin on cells. The same reaction was reported by Iglewski *et al.* (1975) as the mode of action of *P. aeruginosa* toxin A. In both cases the substrate is elongation factor-2 (EF-2). Toxin catalyzed ADP-ribosylation was also reported for cholera toxin and exotoxin of *Bordetella pertussis*, the substrate being regulatory components of membrane adenylate cyclase (Moss and Vaughan, 1979; Katada and Ui, 1982).

Our studies demonstrate that the molecular action of *P. pseudomallei* exotoxin also involves ADP-ribosylation. However, unlike *P. aeruginosa* toxin A which ADP-ribosylates in its intact form, the *P. pseudomallei* exotoxin seems to require a cleavage step for its enzymatic activity. In this way the *P. pseudomallei* exotoxin behaves in a similar manner to the diphtheria toxin whereby its transferase activity is only evident after limited proteolysis. Fragment A of diphtheria toxin and *P. aeruginosa* toxin A inhibit protein synthesis elongation step by catalyzing the transfer of ADP-ribose moiety of NAD^+ onto EF-2 thus leading to the inability of ADP-ribosylated EF-2 to translocate ribosomes along mRNA (Middlebrook and Dorland, 1984). In the case of *P. pseudomallei* exotoxin further study is necessary to understand the nature of the cleavage products.

The striking similarity of the molecular action of *P. pseudomallei* exotoxin with the diphtheria group and *P. aeruginosa* toxin A was not surprising. Perhaps as suggested by Iglewski and Kabat (1975) : NAD^+ dependent ADP-ribosylation of various target proteins is a common widespread general mechanism of bacterial action holds its truth. As to whether the action of the *P. pseudomallei* exotoxin involving ADP-ribosylation of EF-2 proves to be the primary biochemical lesion in the septicaemic form of melioidosis awaits further investigation.

ACKNOWLEDGEMENTS

We wish to acknowledge the support of IRPA grant (1-07-03-012) from the Malaysian Government. We also like to thank En. Ahmad Mansor for technical assistance and Pn. Siti Hajar for typing the manuscript.

REFERENCES

Clemens, M.J. (1984) Translation of Eukaryotic messenger RNA in cell-free extracts. In: *Transcription and Translation a practical approach*, edited by B.D. Hames and S.J. Higgins, pp.231-269. Oxford: IRL Press.

Dannenberg, A.M. and Scott, E.M. (1958) Melioidosis pathogenesis and immunity in mice and hamsters. *J. Exp. Med.* **107**, 153.

Hayaishi, O. and Ueda, K. (1979) Poly (ADP) ribose and ADP-ribosylation of proteins. *Annu. Rev. Biochem.* **46**, 95–116.

Iglewski, B.H. and Kabat, D. (1975) NAD^+-dependent inhibition of protein synthesis of *P. aeruginosa* toxin. *Proc. Natl. Acad. Sci. USA.* **72**, 2284–2288.

Katada, T. and Ui, M. (1982) Direct modification of the membrane adenylate cyclase system by islet-activating protein due to ADP-ribosylation of a membrane protein. *Proc. Natl. Acad. Sci. USA* **79**, 3129-3133.

Maheswaran, P., Mohamed, R., Noor Embi, Omar, O., Senafi, S., Nyonya, R. and Ismail, G. (1987) Cytotoxicity studies of *P. pseudomallei* exotoxins on cells in culture. *Proc. Mal. Biochemical Soc. Conf.* **13**, 254–260.

Middlebrook, J.L. and Dorland, R.B. (1984) Bacterial toxins: Cellular mechanisms in action. *Microbiol. Rev.* **84**, 199–221.

Mohamed, R., Nathan, S., Noor Embi and Ismail, G. (1989) Inhibition of protein synthesis in macrophages by the exotoxin on *P. pseudomallei. Microbiol. Immunol.* **33**, 811-820.

Mohamed, R., Periasamy, M., Hong, G.K., Ismail, G. and Noor Embi (1989) Inhibition of DNA synthesis in cultured cells by the exotoxin of *P. pseudomallei. Sains Malaysiana.* (in press).

Moss, J. and Vaughan, M. (1979) Activation of adenylate cyclase by choleragen. *Ann. Rev. Biochem.* **48**, 581–600.

Nigg, C., Heckley, R.J. and Colling, M. (1955) Toxin produced by *Malleomyces pseudomallei. Proc. Soc. for Expt. Biol. & Med.* **89**.

Nyonya, R., Ismail, G., Noor Embi and Omar, O. (1986) Protection studies using whole cells and partially purified toxic material of *P. pseudomallei. Malays. Appl. Biol.* **15**, **105–111**.

Smith, C.J., Allen, J.C., Noor Embi, Othman Omar, Razak, N. and Ismail, G. (1987) Human melioidosis: an emerging medical problem. *Mircen J.* **3**, 343–366.

Wasiman, I., Noor Embi, Mohamed, R., Omar, O., Razak, N. and Ismail, G. (1987) The purification of *P. pseudomallei* exotoxin. *Proc. Mal. Biochemical Soc. Conf.* **13**, 241-247.

Whitmore, A. and Krishnaswami, C.S. (1912) An account of the discovery of a hitherto undescribed infective disease occurring among the population of Ranggon. *Indian Medical Gazette* **97**, 262–267.

15 TWO NATURALLY OCCURRING TOXINS CAUSING STOCK LOSSES

PETER B. OELRICHS AND DUDLEY H. WILLIAMS

INTRODUCTION

Serious losses of livestock in Queensland due to poisoning by naturally occurring toxic substances have been reported since early settlement. Most of the cases are caused by poisonous plants or microbial toxins in plants. Overall estimates of economic losses to the Pastoral and Agricultural Industries in Australia have been made by Culvenor (1984) and McKenzie (1984) and are of the order of $70–80 million per annum. This economic loss appears in many forms. In addition to the direct loss of the animals, many poisonous plants cause a significant decrease in reproduction efficiency and growth rate. Also animals with birth defects caused by poisonous plants may be worthless (Nielsen *et al.*, 1989). There are many indirect costs associated with plant poisoning. Additional management procedures, for example fencing, may be necessary, and in some cases productive land is not utilized because it is considered hazardous to stock. Another most important aspect of these problems, not often considered, is the accumulation of plant toxins in animal tissues used for human consumption. Little is known about natural toxicants in this context partly because of the large number of potential toxins and partly because of their limited distribution in plants (Hegarty, 1989a).

The present article gives a brief review of the work which lead to the structure elucidation of moroidin and lophyrotomin, two toxic peptides obtained from natural sources in Queensland.

Intoxication of animals by either toxic peptide is unusual in that it does not follow the familiar pathway which leads to death of the animal from plant poisoning. With moroidin, victims are poisoned by touching needles on *Dendrocnide moroides* (Gympie Stinging Tree) leaves which inject the potent stinging toxin, and with lophyrotomin, poisoning occurs when animals eat the toxic *Lophyrotoma interrupta* sawfly larvae found on *Eucalyptus melanophloia* (Silverleafed Ironbark) trees in a well defined area of Western Queensland. Recently lophyrotomin was isolated from a different cattle killing sawfly species *Arge pullata* found on *Betula pendula* trees (Silver Birch) growing in Denmark. Each of these problems will be discussed separately.

MOROIDIN

The severe stinging properties of the three *Dendrocnide* species following cutaneous contact have been recognized for over a century. *D. excelsa* and *D. photinophylla* are trees growing to 35 m and occur in rainforests of the area from the Victorian border to the Atherton Tableland in North Queensland. *D. moroides* is a branched shrub growing to 4 m commonly found from about Gympie in South Queensland to Cape York peninsula in the far north (Figure 15.1). It also extends outside Australia to the southern Moluccas, but apparently is not recorded in New Guinea (Everist, 1974). As early as 1887, there were reports of horses being killed by contact with the plants, and frequently animals which had been stung had to be rested for several days before they could be used again (Bailey *et al.*, 1887). Bancroft (1889) observed the long duration of the sting which lasts several weeks in some cases, and is particularly noticeable when the area is wet. Farmers and timber workers are the main sufferers and frequently victims become frantic if badly stung. In some cases unconsciousness has been reported in victims. A human fatality has been reported in New Guinea following the sting of a similar species *D. cordata* (Winkler, 1922). Because *D. moroides* gave the most vicious sting both in intensity and duration it was considered the most suitable for use in our investigations to isolate and identify the active stinging compound(s).

The sting from the plants occurs when the skin is pierced by long needle-like hairs found on the leaves and green stems of the plants. The hairs are poison reservoirs with siliculated tips at the upper end. The tips break off easily making each hair into a sharp

Figure 15.1 *Dendrocnide (Laportea) moroides* fruit and leaf.

pointed cannula through which the poison liquid is ejected into the injured skin (Schildknecht, 1981). The results of the cutaneous contact have been described by Webb (1948), Francis (1955) and Robertson *et al.* (1957). Very often pain can last for several days, and even when it disappears the affected area may be extremely painful during cold weather or if rubbed or if wet with cold water. In addition to the cutaneous reaction following contact with *D. moroides* stinging hairs there is a severe sneezing reaction produced when the shrub is contacted or approached from the windward side (Schildknecht, 1981). This phenomenon could present an additional hazard to asthma sufferers.

Early work by Petrie (1906) claimed formic and acetic acids were present in the stinging hairs of *Dendrocnide* species in sufficient amount to cause the severe stings. Later Rechter (1949) observed the extracts of Dendrocnide species leaves showed pharmacological reactions similar to acetylcholine and histamine. Robertson *et al.* (1957) demonstrated acetylcholine like, histamine like and 5-hydroxytryptamine like substances in extracts of stinging hairs, but intradermal injection of the dialysed extracts of stinging hairs free of these compounds produced responses similar to extracts of the hairs alone on injection. They concluded the essential pain producing material appeared to be a stable, non dialysable substance resistant to heat, neutral in reaction and not attacked by proteolytic enzymes.

Oelrichs *et al.* (1970) made extracts of large quantities of milled air dried leaf rather than individual hairs to obtain sufficient toxin for testing and chemical work. The presence of the toxin was established by concurrent toxicity tests as outlined by Robertson *et al.* (1987). The yield of toxin isolated using this method was 2–3 mg from 100 g of dried *D. moroides*, and this produced a similar biological reaction to the previous extracts of stinging hairs. No significant structural work was carried out on this compound because the amount isolated was too small. An improved method (Leung *et al.*, 1986) using frozen leaves and stalk of *D. moroides* yielded 10–12 mg of toxin termed moroidin from about the same amount of plant.

By a combination of chemical, ^{1}H and ^{13}C NMR and mass spectrometric studies, two alternative structures for moroidin were elucidated. Moroidin was shown to be a bicyclic octapeptide containing an unusual C-N linkage between the tryptophan and histidine residues. The final solution to the structure was devised by Kahn *et al.* (1989) using a combination of NMR and molecular modelling techniques. The structure of moroidin is shown to be composed entirely of L-amino acids. The unusual C-N linkage between the tryptophan and histidine residues has been found to be between C_2 of tryptophan and N_1 of the imidazole moiety of histidine. The following structure (Figure 15.2) is thus correct for moroidin.

Biological testing with pure moroidin gave less immediate response than some impure extracts, but moroidin appeared to be responsible for the reported long duration of the sting. A stinging sensation in the area injected with moroidin could be felt even after several months, and this could be precipitated by a change of temperature.

The role of moroidin and other stinging compounds such as the biogenic amines in *D. moroides* appears to be defence. Their pharmacological properties and occurrence in defence organs clearly demonstrates their defensive function. The sneeze producing property of *D. moroides* which is thought to come from smaller gland cells on the leaves and stems (Schildknecht, 1981) has not yet been fully investigated. The mechanism of the effects of the stinging compounds when injected intradermally is not

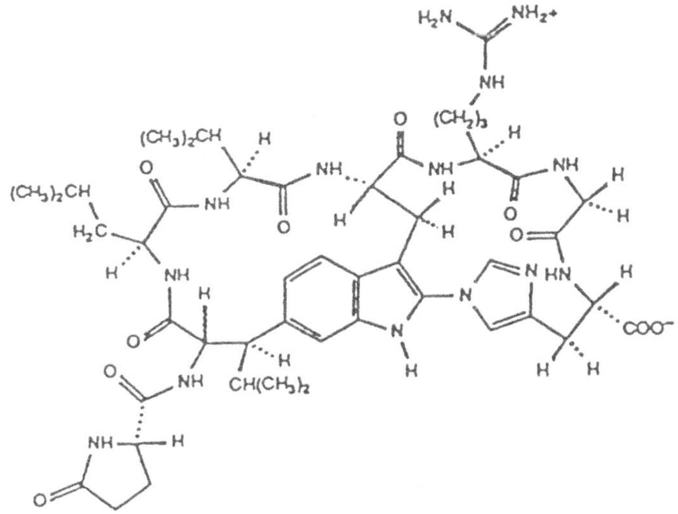

Figure 15.2 S–(Cαβ[s]Leu), S–(CαTrp), C_2–N_1 imidazole linkage in moroidin.

known. The pain sensation occurs 15–30 s after injection, and the fact that the latency is constantly observed suggests an activation takes place in the skin.

LOPHYROTOMIN

The sawfly poisoning of sheep and cattle by ingestion of the larvae of the sawfly *Lophyrotoma interrupta* occurs mainly in the Maranoa, Warrego and Leichardt districts of Queensland (Roberts, 1932). It is here large forests of *E. melanophloia* the principal host tree of the larvae are found. The disorder usually occurs in late winter or early spring if there is a heavy infestation of the trees. During this period larvae descend to the ground to pupate, and often large numbers congregate at the bases of trees where they form large mounds and die. It is a remarkable feature of the sawfly poisoning disorder that animals, notably cattle, eat such material and in addition develop a considerable liking for it. Animals that have become addicted to eating larvae voraciously consume both dead and living larvae, and have been seen fighting one another in attempts to obtain more of the abnormal material which in short time causes their death (Callow, 1955). Large numbers of cattle can be killed, and the estimated yearly cost to the cattle industry is $1.0 million (McKenzie, 1984). The clinical signs and pathology of poisoned animals have been reviewed by McKenzie *et al.* (1985). The main site of action of the toxin appears to be the liver, and this shows areas of fatty infiltration and necrosis. Extensive jaundice because of the liver damage is often seen.

Leonard (1972) first reported the isolation of a toxic factor from sawfly larvae. An improved method for the isolation of the toxin was reported by Oelrichs *et al.* (1977). They reported the isolation of an essentially pure toxin with LD_{50} of 2 mg/kg. On the basis of evidence from chemical methods, NMR and mass spectrometry a tentative structure for the new toxic octapeptide termed lophyrotomin was proposed.The remaining problems in the structure determination were elucidated by NMR studies, sequence studies and molecular weight determination by mass spectrometry, FAB mass spectrometry, gas liquid chromatography on a chiral column and chemical methods (Williams *et al.*, 1983). The structure of the new octapeptide lophyrotomin was shown to be:

C_6H_5CO-(D)-Ala-(D)-Phe-(L)-Val-(L)-Ile-(D)-Asp-(L)-Asp-(D)-Glu-(L)-Gln

The presence of four D-amino acids in a peptide isolated from an animal is, so far as we are aware, unique and suggested the peptide was generated by a non-ribosomal biosynthetic pathway in micro organisms occurring in larvae. In the course of investigation of this possibility a fungus identified as *Beauveria bassiana* was isolated from living *L. interrupta* larvae and shown to biosynthesize beauvericin, a cyclic peptide (Peeters, 1983).

Cell free extracts prepared from lyophilized mycelium of *B. bassiana* isolated from *L. interrupta* were able to synthesize beauvericin *in vitro.* The beauvericin synthesizing activity could be enriched about 12-fold by fractional ammonium sulfate precipitation. Beauvericin (Figure 15.3) is a cyclo-depsipeptide with insecticidal properties, and may be regarded as a homologue of the ionophoric antibiotics enniatins A-C. These micro organisms did not synthesize lophyrotomin or any related compound.

The European birch sawfly *Arge pullata* was first reported in Denmark in 1974 and was thought to have been introduced from the Baltic areas. The sawfly is now widespread (Martin, 1974). In September 1984 on the Danish Island of Sjaelland heavy losses occurred in sheep which had been moved to new pasture. The pasture contained many thickets of birch trees which were heavily infested with *A. pullata* larvae. Thamsborg *et al.* (1987) made observations of field cases at Sjaelland, and carried out laboratory experiments with sheep to determine the toxicity of *A. pullata*. The marked similarity of these cases to poisoning by the larvae of *L. interrupta* (Figure 15.4) in Queensland prompted our investigations. The toxin was isolated from *A. pullata* sawfly and shown by chemical and spectrometric evidence to be identical to lophyrotomin (Kannan *et al.*, 1988).

The work of Eisner *et al.* (1974) clearly demonstrates the protective role of an oily resin discharged by the sawfly *Neodiprion sertifer*. This resin is found in the host tree *Pinus sylvestris*, and is sequestered in diverticular pouches by the larvae upon feeding. *L. interrupta* larvae are known to discharge a similar highly irritant oily fluid which was shown to contain a high concentration of lophyrotomin. In this case no detectable amount of lophyrotomin was found in the host tree *E. melanophloia* (unpublished work). The available evidence therefore suggests lophyrotomin has a defensive role and is either biosynthesized by the larvae or by another as yet unidentified micro organism(s). The fact that no

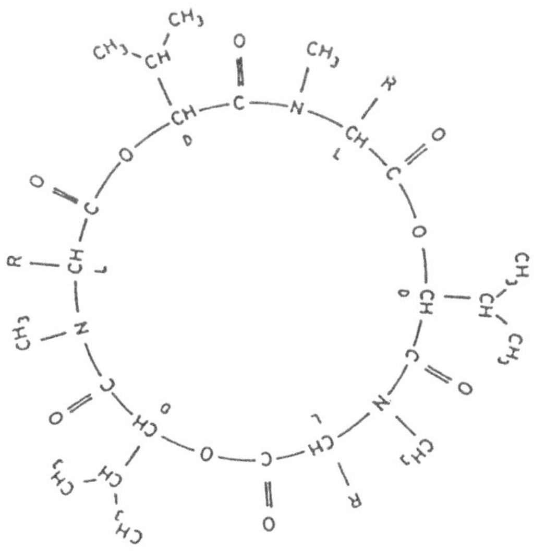

Figure 15.3 Beauvericin, R = $-CH_2C_6H_5$; Enniatin A, R = $-CH(CH_3)CH_2CH_3$; Enniatin B, R = $-CH(CH_3)_2$; Enniatin C, R = $-CH_2CH(CH_3)_2$. (Reprinted with permission from *The Journal of Antibiotics* **36**, 1762–6, 1983, Japan Antibiotics Research Association.)

Figure 15.4 *Lophyrotoma interrupta* sawfly larvae.

B. bassiana or related micro organisms have been detected in *A. pullata* larvae (Figure 15.5) gives weight to the argument that lophyrotomin is biosynthesized by the larvae *per se* and not by micro organisms in the larvae.

Another possibility is that lophyrotomin may act as an antimicrobial agent in the larvae (Dr Hans von Döhren, Berlin, personal comm. 1983). In this case it may act to control the growth of some organisms, so that others essential for its survival can survive. It is interesting to speculate that these organisms that are controlled may include viruses. A gut cell-infecting nuclear polyhedrosis virus has been detected in a related gut regurgitating sawfly (Dr P.F. Entwistle, Oxford personal comm. 1987).

A large number of natural peptides are known to have biological activity and some are toxic to animals. Amongst these are the cyclopeptides which include the well known hepatotoxins from the *Amanita* species of mushrooms (Weiland *et al.*, 1972). The symptoms of intoxication, pathology and approximate LD_{50} of lophyrotomin have been described (McKenzie *et al.*, 1985) and

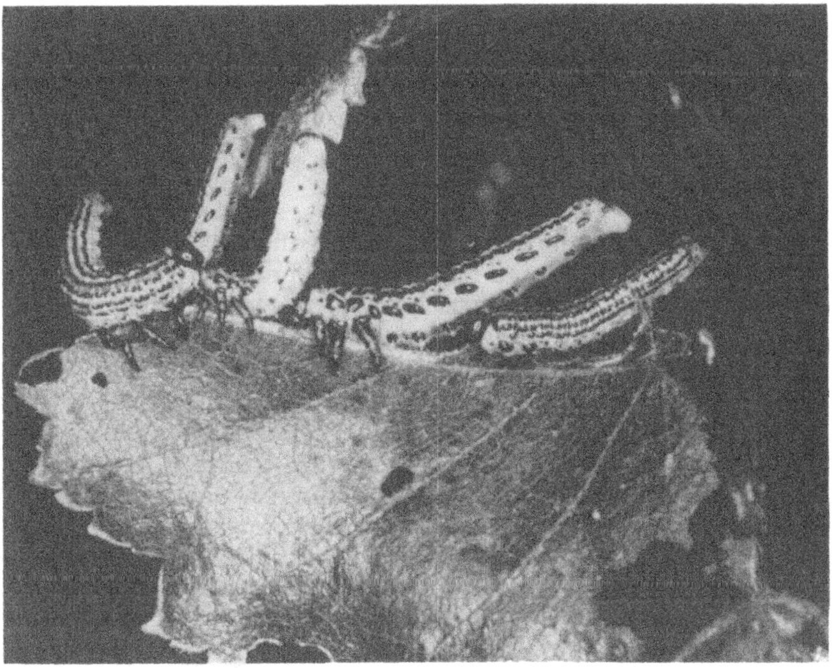

Figure 15.5 *Arge pullata* sawfly larvae.

resemble those of the hepatotoxins from *Amanita* species. Oelrichs (1982) suggested silybin, the antidote used successfully in *Amanita* poisoning of human beings, may be effective if administered early in the intoxication of cattle with lophyrotomin or as a propylactic.

Although the site of action of the toxin is not known, it is noteworthy that the N-terminal region could be bound to the cell lipid membrane, and the C-terminal region free to bind the metal ions (such as Ca^{++}) and have an effect on cell membranes. *Amanita* toxins are known to produce deformation of hepatocyte membranes, but a similar effect was not observed in one *in vitro* experiment with lophyrotomin (Dr M.T. Runnegar, personal comm.). In overcoming *Leucaena* toxicity caused by the alkaloid mimosine, Jones (1984) showed the beneficial effects of transferring mimosine degrading rumen microbes from one animal to another. This method offers possibilities for overcoming other plant related toxicity problems. In view of the enormous variation in the ability of bacteria to attack a wide range of substrates, and the molecular biological methods now available to transfer detoxifying ability, it is possible for a solution to the sawfly poisoning problem (and many others) to be found.

CONCLUSIONS

A most important aspect of this work is the great potential many of these toxins in plants have for use in medicine. Through the ages plants or plant extracts have been used for treatment of a variety of human ailments and disorders, and today many of our commonly used drugs in medicine are plant products or their derivatives. To take just two examples of current interest. Firstly, castanospermine a new alkaloid from *Castanospermum australe*, which is a potent inhibitor of several α- and β-glucosidases. A major discovery was that castanospermine was active against HIV *in vitro* (Tyms *et al.*, 1987) and recent work has shown its 6-butyryl derivative is 20 times more active than the parent alkaloid (Hegarty, 1989b). Second, several plant toxins (for example ricin, a toxalbumin) have been used successfully in a study of the selective destruction of tumor cells by linking these cytotoxins to monoclonal antibodies (Pietersz, 1989). The structures and biological action of the more important naturally occurring compounds used for medical purposes are known. However although a very large number of other compounds with biological activity have been isolated and identified in many cases their mode of action has not been determined. The fact that in the rain forests of the world there still exist plant species that have not yet been identified suggests a great potential resource for the future when the toxins and other biologically active compounds from these can be fully investigated.

REFERENCES

Bailey, F.M. and Gordon, P.L. (1887) *Plants reputed poisonous and injurious to stock*. Brisbane: J.C. Beal, Govt. Printer.

Bancroft, T.L. (1889) Intercolonial Medical Congress, Australian, Transactions, p. 927.

Callow, L.C. (1955) Sawfly poisoning in cattle. *Qld. Agric. J.* **81**, 155–161.

Culvenor, C.C.J. (1984) Economic loss due to poisonous plants in Australia. In: *Plant Toxicology*, Proceedings of the Australia-U.S.A. Poisonous Plants Symposium, (Brisbane 1984), pp. 3–13.

Eisner, T., Johnesse, J.S., Carrel, J., Hendry, L.B. and Meinwold, J. (1974) Defensive use by an insect of a plant resin. *Science* **184**, 996–999.
Everist, S.L. (1974) *Poisonous Plants of Australia*, 730 pp. London, Sydney: Angus and Robertson.
Francis, W.D. (1955) *Australian Stinging Trees*, C.J. White Memorial Lecture, Brisbane.
Hegarty, M.P. (1989a) The hepatotoxicity of indospicine in dogs. In: *Proceedings of the 3rd International Symposium on Poisonous Plants*, Logan Utah (in press).
Hegarty, M.P. (1989b) In: *A Report to the Queensland Poisonous Plants Committee*, (Nov. 1989).
Jones, R.J. (1984) Leucaena toxicity and the ruminal degradation of mimosine. In: *Plant Toxicology*, Proceedings of the Australia-U.S.A. Poisonous Plants Symposium, (Brisbane 1984), pp. 111–119.
Kahn, S.D., Booth, P.M. and Williams, D.H. (1989) Computer-assisted structure determination. Structure of the peptide moroidin from *Laportea moroides*. *J. Org. Chem.* **54**, 1901–4.
Kannan, R., Oelrichs, P.B., Thamsborg, S.M. and Williams, D.H. (1988) Identification of the octapeptide lophyrotomin in the European birch sawfly (*Arge pullata*). *Toxicon* **26**, 224–226.
Leonard, G.J. (1972) The isolation of a toxic factor from sawfly (*Lophyrotoma interrupta* Klug) *Toxicon* **10**, 597.
Leung, T-W.C., Williams, D.H., Barna, J.C.J., Foti, S., and Oelrichs, P.B. (1986) Structural studies on the peptide moroidin from *Laportea moroides*. *Tetrahedron* **42**, 3333–3348.
McKenzie, R.A. (1984) The cost of cattle deaths from toxic plants in Queensland 1972–1981. In: *Plant Toxicology*, Proceedings of the Australia-U.S.A. Poisonous Plants Symposium, (Brisbane 1984), p. 14.
McKenzie, R.A., Dunster, P.J., Twist, J.O., Dimmock, C.K., Oelrichs, P.B., Rogers, R.J. and Reichmann, K.G. (1985) The toxicity of sawfly larvae (*Lophyrotoma interrupta*) to cattle. *Qld Dept. Primary Ind. Bull*, Q B85001.
Martin, O. (1974) Kaempeangreb pa birk af ny dansk bladhveps. *Naturnes Verden* **5**, 188.
Neilsen, D.B. and James, L.F. (1989) Economic loss due to poisonous plants in U.S.A. In: *Proceedings of the 3rd International Symposium on Poisonous Plants* at Logan, Utah (in press).
Oelrichs, P.B. (1982) Sawfly poisoning in cattle. *Qld Agric. J.* **108**, 110.
Oelrichs, P.B. and Robertson, P.A. (1970) Purification of pain-producing substances from *Dendrocnide (Laportea) moroides*. *Toxicon* **8**, 89–90.
Oelrichs, P.B., Vallely, P.J., MacLeod, J.K., Cable, J., Kiely, D.E. and Summons, R.E. (1977) Lophyrotomin, a new toxic octapeptide from the larvae of sawfly *Lophyrotoma interrupta*. *Lloydia* **40**, 209–214.
Peeters, H., Zocher, R., Madry, N., Oelrichs, P.B., Kleinkauf, H. and Kraepelin, G. (1983) Cell-free synthesis of the depsipeptide beauvericin. *J. of Antibiotics* **36**, 1762–1766.
Petrie, J.M. (1906) *Proc. of the Linnean Soc. of N.S.W.* **31**, 530–545.
Pietersz, G.A. and McKenzie, I. (1989) Chapter 8, this volume.
Rechter, B. (1949) M. Sc. Thesis, University of Melbourne.
Roberts, F.H.S. (1932) Sawfly poisoning in cattle. *Qld. Agric. J.* **37**, 41.

Robertson, P.A., and MacFarlane, W.V. (1957) Pain-producing substances from the stinging bush *Laportea moroides*. *Aust. J. Exp. Biol. and Med. Sci.* **35**, 381–394.

Schildknecht, H. (1981) Irritant and defence substances of higher plants. *Angew. Chem. Int. Ed. Engl.* **20**, pp. 164–184.

Thamsborg, S.M., Jorgensen, R.J. and Brummerstedt, E. (1987) Sawfly poisoning in sheep and goats. *The Veterinary Record* September 12, 253–255.

Tyms, A.S., Berrie, E.M., Ryder, T.A., Nash, R.J., Hegarty, M.P., Taylor, O.L., Mobberley, M.A., Davis, J.M., Bell, E.A., Jeffries, D.J., Taylor-Robinson, D. and Fellows, L.E. (1987) Castanospermine and other plant alkaloid inhibitors of glucosidase activity block the growth of HIV. *Lancet*, **ii** (31 October):1025–26.

Webb, L.J. (1948) *CSIRO Bulletin* No **232**, 166.

Weiland, T. and Weiland, O. (1972) The toxic peptides of *Amanita* species. In: *Microbial Toxins*, edited by S. Kadis, A. Ceigler and S.J. Ajl, Vol. VIII, pp. 249–280. London, New York: Academic Press Publishers.

Williams, D.H., Santikarn, S., De Angellis, F., Smith, R.J., Reid, D.G., Oelrichs, P.B. and MacLeod, J.K. (1983) The structure of a toxic octapeptide from the larvae of sawfly. *J. Chem. Soc. Perkin Tran.* **1**, 1869–1878.

Winkler, H. (1922) Die Urticaceen Papuasiens. *Engler's Botanische Jahrbücher* **57** (3–5), 501–509.

16 A JELLYFISH TOXIN

NOR AZILA AND IEKHSAN OTHMAN

INTRODUCTION

The general unpleasant contact with venomous coelenterates has been the main cause of concern since these animals are usually found in the shallow coastal waters which people use for sports and recreation. Jellyfish venoms, contained in the nematocysts embedded in the tentacles, consist of complex and highly concentrated mixtures of proteins and polypeptides which possess enzymatic and a number of physiological activities. The coelenterates cause envenomation which may result in fatal, systemic or local syndromes. These may manifest as a stinging or burning sensation followed by the development of red elevated lesions, tightness of the chest, severe muscular cramps and vomiting.

The toxicity of the jellyfish venom is attributable to proteins capable of disrupting membrane structure and function (Ferlan and Lebez, 1974; Toom and Phillips, 1975; Burnett and Carlton, 1977). The Jellyfish *Catostylus mosaicus*, or commonly referred to as the 'blubber' is most commonly found around the eastern Australian coastline (Southcott, 1987) and the coastal waters of the Malaysian Peninsula.

PREPARATION OF VENOM EXTRACT AND ENZYMIC ACTIVITIES

The crude venom extract (CE) was prepared from the tentacles of *Catostylus mosaicus* as described in Azila and Othman (1990). The crude extract was subjected to acetone precipitation with 2 volumes of cold acetone (67% saturation) and then centrifuged. The precipitate was resuspended in 0.01 M Tris buffer (pH 7.4) containing 0.14 M NaCl and 0.1 mM $CaCl_2$ (Tris-buffered-saline, TBS). The supernatant was rotary-evaporated to remove the acetone and was labelled as acetone-soluble extract (AE). When freeze-dried AE was redissolved in 0.1 M Tris-HC1 buffer (pH 7.4) containing 0.5 M NaCl, a thin suspension of lipid was visible at the top of the solution and this was removed by centrifugation. The clear solution was labelled as S.

The crude extract was devoid of any hydrolytic enzymes normally present in venoms except for phospholipase A, with the specific activity of 135.6 m mol/min/mg extract protein. Acetylcholinesterase, phosphomonoesterase and protease were assayed according to the method described by Tan and Tan (1988a), and phospholipase A activity was measured by the acidimetric assay of Tan and Tan (1988b). The phospholipase A hydrolyzed phospholipids from erythrocyte membranes with the concomitant appearance of their lyso-derivatives and fatty acids (Table 16.1).

Table 16.1 Lipid analysis of rabbit erythrocyte membranes treated with crude venom extract

Lipid	*Control*	*Treated*	*% change*
	(mg. lipid/mg. membrane protein)		
PE	0.64 + 0.05	0.43 + 0.08	–32.4
PC	0.74 + 0.1	0.55 + 0.09	–25.7
SM	0.44 + 0.06	0.47 + 0.06	not significant
LPE	nd	0.49 + 0.02	+ (*)
LPC	nd	0.23 + 0.06	+ (*)
FFA	nd	2.46 + 0.3	+ (*)
Chol.	1.1 + 0.3	1.25 + 0.1	not significant

Abbreviations:
PE – phosphatidylethanolamine, PC – phosphatidylcholine, SM – sphingomyelin, LPE – LysoPE, LPC – lysoPC, FFA – free fatty acid, Chol. – cholesterol.

(*) Appearance of the lipid upon treatment with crude venom.
(–) Indicate decrease of lipid content compared to controls.
nd Not detectable

2 ml of 10% erythrocyte suspension was incubated with an equal volume of crude extract (5.4 mg protein) at 28°C. After 30 min an aliquot was taken out and diluted with TBS to give a final concentration 1% erythrocyte suspension, to estimate the extent of haemolysis. The remaining mixture was extracted for lipid (Yagihara *et al.*, 1973) and the various lipids were separated by TLC. Lipid content was determined by the simple charring method of Marsh *et al.* (1966). Phospholipid was estimated by measuring the phosphate content using the method of Bartlett (1959).

HAEMOLYTIC ACTIVITY

The haemolytic activity was estimated following the method of Jorgensen *et al.*, (1982), with slight modifications. Fresh erythrocytes from the rabbit were washed with TBS. A final 1% erythrocyte solution was prepared for use in the haemolytic determination. The amount of haemoglobin released from the erythrocytes was used as an index to measure haemolysis. One haemolytic unit is defined as the amount of protein which brought about 50% haemolysis in 30 min at 28°C.

The crude extract caused lysis of erythrocyte membranes from different mammalian species with different potencies, where 20 mg of venom protein was shown to cause approximately 40%, 50% and 90% haemolysis of human, rabbit and rat erythrocytes, respectively (Figure 16.1). The haemolytic activity was not linear but increased in a sigmoidal manner with increasing concentra-

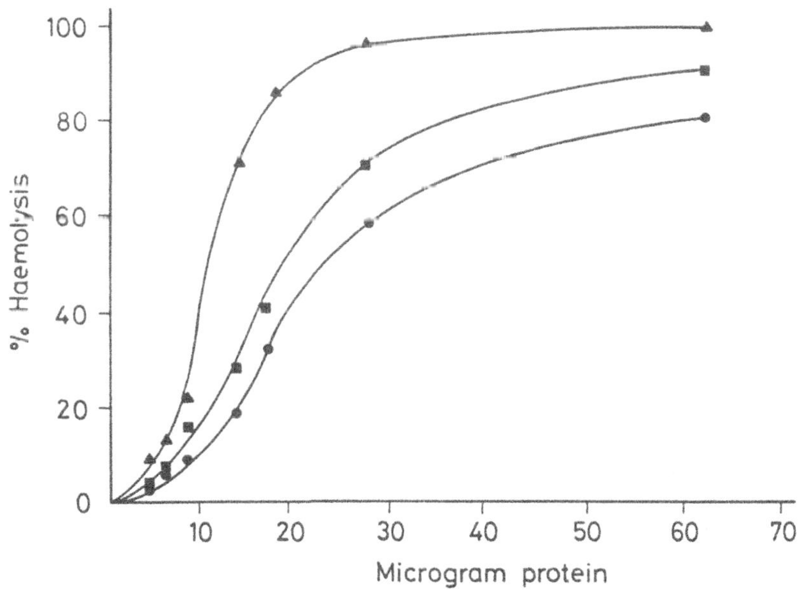

Figure 16.1 Suspectibility of rat (▲), rabbit (■) and human (●) erythrocytes to increasing concentrations of crude venom. A 1% erythrocyte (100 ml) suspension was incubated with crude venom extract (100 ml) at 20°C for 30 min. The mixture was then made up to 1 ml with TBS followed by centrifugation at 10 000 g. The absorbance of the supernatant was read at 540 nm. Control for 0% and 100% haemolysis was obtained by incubation of 1% erythrocyte with 0.9 ml TBS and ammonium hydroxide, respectively.

tions of the crude protein extract. The haemolytic component was acetone soluble since the precipitated fraction was devoid of any haemolytic activity while the supernatant fraction showed a 24 fold increase in the haemolytic activity. The haemolytic activity of crude extract was 50 U/mg protein while that of the partially purified fraction after acetone precipitation was 1204 U/mg protein.

The haemolytic activity of the crude extract was quite stable to heat, since heating the extract at 100°C for one hour only reduced its activity to approximately 60%. However, after extraction with acetone and the removal of the lipid component there was a loss in the stability of the haemolytic component (Figure 16.2).

Fractionation of the solubilized fraction (S) on Sephacryl S-300 gave two peaks of haemolytic activity with peak I exhibiting a much higher activity than peak II (Figure 16.3).

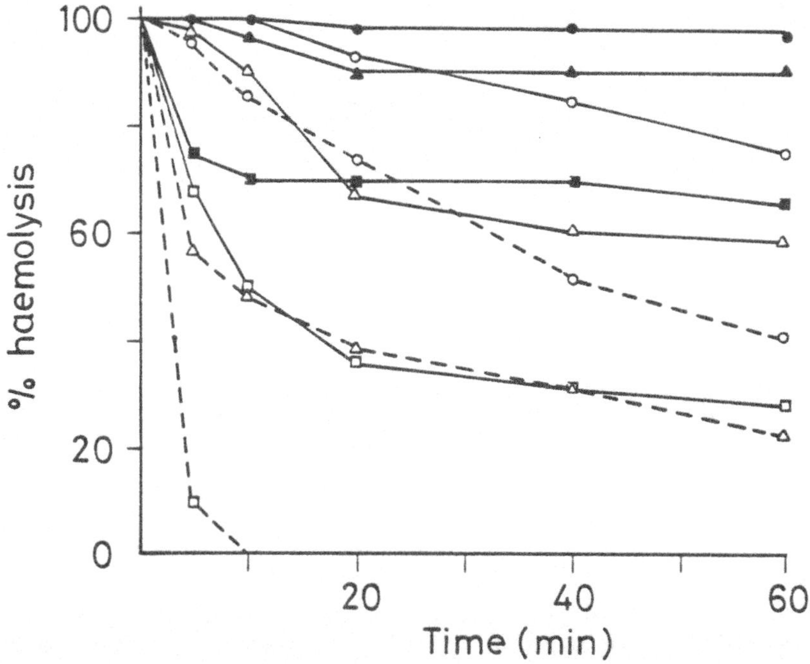

Figure 16.2 Effect of heat pretreatment on haemolytic activity of samples at various purification steps. (o)—40°; (Δ)—60° and (□)—100°C. Closed symbol represents CE, open symbols with full lines for AE and open symbol with broken lines for S.

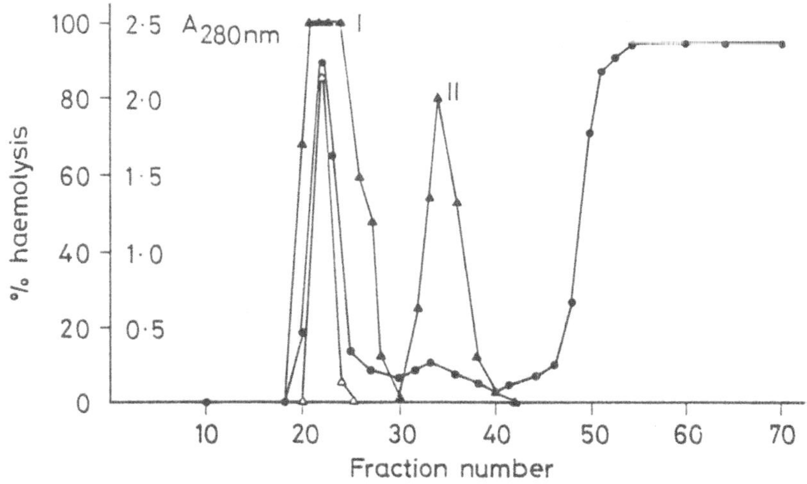

Figure 16.3 Sephacryl S–300 chromatography of the acetone soluble extract. (▲–▲), percentage haemolysis produced by the fractions; (Δ–Δ), percentage haemolysis with 40 fold dilution of the fractions and (●-●) absorbance at 280 nm.

HYDROLYSIS OF PHOSPHOLIPIDS ON ERYTHROCYTE MEMBRANE

Time course studies (Azila and Othman, 1990) had shown that although the haemolytic profile was almost parallel to that of hydrolysis of the erythrocyte membrane phospholipids, the rate of increase of the latter was not as fast as that of haemolysis. Hydrolysis of membrane phospholipid was also quite extensive, with approximately 60% of PC, 50% SM and 30% of PE hydrolyzed after 20 min, resulting in the generation of lysophospholipids. Lysophospholipids had been shown to be lysogenic (Lucy, 1970), which may contribute to haemolysis. However, hydrolysis of erythrocyte membrane phospholipids by the action of phospholipase A alone, had been shown not to cause haemolysis (Roelofsen *et al.*, 1971; Lankish and Vogt, 1972). It was reported that haemolysis will only occur if sublytic concentrations of lysins were included. This cooperative effect between phospholipase A activity and haemolysins may explain the sigmoidal dose-response curve obtained.

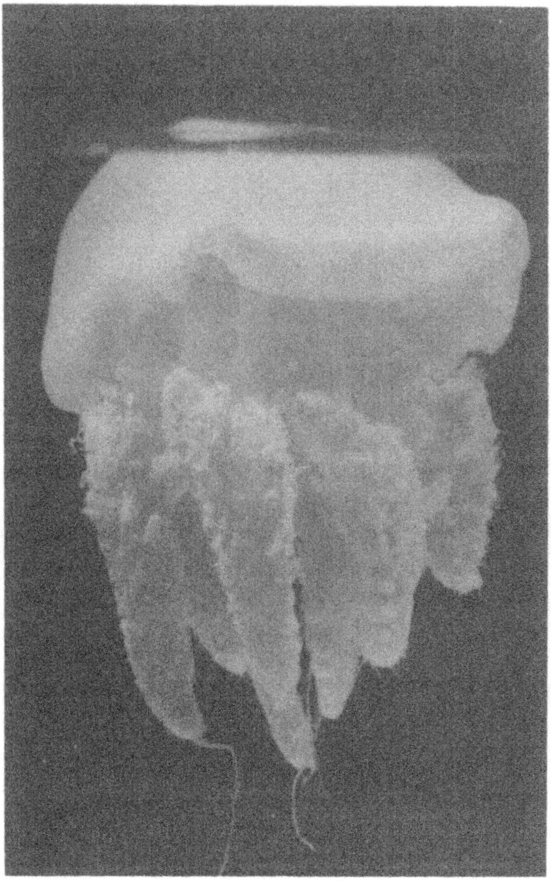

Figure 16.4 The jellyfish *Catostylus mosaicus.*

CONCLUDING REMARKS AND CURRENT PROGRESS

At present we are not able to say with certainty whether the haemolytic activity present in this extract is due purely to a single component or the cooperation of several components. The cooperativity is suggested by the sigmoidal dose-response curve. The phospholipase A activity detected in the extract could contribute to haemolysis induced by haemolytic protein(s) present. On the basis that the haemolysin in the crude extract could not be precipitated even with 80% saturation of acetone and was stable to heat, we would suggest that the protein involved is not a simple protein, but rather a complex protein.

Up till now we have not been able to purify the haemolytic component(s). Proteins exhibiting haemolytic activity could be precipitated with ammonium sulphate but the activity was recovered over a broad spectrum of ammonium sulphate saturation, i.e. from 10 up to 80% saturation. Gel filtration chromatography of the four salt cuts indicated three forms of the haemolytic component. Ion-exchange chromatography of the acetone-soluble extract showed that cationic/neutral and anionic haemolytic components were recovered. Three anionic haemolytic peaks were detected (in publication). The crude venom extract and the acetone-soluble extract also exhibited oedema-inducing and haemorrhagic activity. Fractions of the venom from the anionic exchanger exhibited increased potency in haemolytic, haemorrhagic and oedema-inducing activities. Current studies on nerve-muscle preparations have indicated that the venom extract is also neurotoxic and myotoxic.

ACKNOWLEDGEMENTS

We thank Mr V. Devadason and Ms Azlila Yunus for their technical assistance. Their work was supported by Vote PJP 84/89 a grant obtained from the University of Malaya.

REFERENCES

Azila, N. and Othman, I. (1990) Haemolytic and phospholipase A activities of the tentacle extract of *Catostylus mosaicus*. *Biochem. Int.* **20**, 291–229.

Bartlett, G.R. (1959) Phosphorus assay in column chromatography. *J. Biol. Chem.* **234**, 466–468.

Burnett, J.W. and Carlton, C.J. (1977) The chemistry and toxicology of some venomous pelagic coelentrates. *Toxicon* **15**, 177–196.

Ferlan, I. and Lebez, D. (1974) Equinatoxin, a lethal protein from *Actinia equina*-1. Purification and characterization. *Toxicon* **12**, 57–6.1

Jorgensen, S.E., Mulcahy, P.F., Wu, G.K. and Louis, C.F. (1983) Calcium accumulation in human and sheep erythrocytes that is induced by *Escherischia coli* haemolysin. *Toxicon* **21**, 717–727.

Lankish, P.G. and Vogt, W. (1972) Direct haemolytic activity of phospholipase A. *Biochim. Biophys. Acta* **270**, 241–247.

Lucy, J.A. (1970) The Fusion of Biological Membranes *Nature* **227**, 814–817.

Marsh, J.B. and Wienstein, D.B. (1966) Simple charring method for determination of lipids. *J. Lip. Res.* **7**, 574–576.

Roelofsen, B., Zwaal, R.F.A., Comfurius, P., Woodward, C.B. and Van Deenan, L.L.M. (1971) Action of pure phospholipase A_2 and phospholipase C on human erythrocytes and ghosts. *Biochim. Biophys. Acta* **214**, 925–929.

Southcott, R. (1987) Jellyfish of Medical Importance. In: *Toxic Plants and Animals*, edited by Covacevich, J., Davie, P. and Pearn, J. , pp. 92–97. Brisbane: Queensland Museum.

Tan, N.H. and Tan, C.S. (1988a) A comparative study of cobra (Naja) venom enzymes. *Comp. Biochem. Physiol.* **90B**, 745–750.

Tan, N.H. and Tan, C.S. (1988b) Acidimetric Assay for Phospholipase A using egg yolk suspension as substrate. *Anal. Biochem.* **170**, 282–288.

Toom, P.M. and Phillips, T.D. (1975) Effects of purified components of jellyfish toxin (*Stomolophus meleagris*) on active sodium transport. *Toxicon* **13**, 261–271.

Yagihara, Y., Bleasdale, J.E. and Hawthorne, J.N. (1973) Effects of acetylcholine on the incorporation of [^{32}P]-orthophosphate *in vitro* into the phospholipids of subsynaptosomal membranes from guinea-pig brain. *J. Neurochem.* **21**, 173–190.

17 SNAKE VENOM—HAEMOSTATIC COMPONENTS

ALAN WHITAKER AND PAUL MASCI

INTRODUCTION

Haemostatically active components are widely distributed among terrestrial venomous snakes, both Elapidae and Crotalidae. Envenomation by such snakes can result in significant effects mediated by haemorrhage, thrombosis or both. The responsible venom components are mainly proteolytic enzymes and proteinase inhibitors, which can affect blood vessels directly, influence platelet function, or interact directly with the coagulation, fibrinolysis and/or complement cascades. These components are of theoretical interest and some are of medical importance as laboratory probes or potential therapeutic agents. We are studying Australian snake venoms, which have always received international attention and have figured prominently in haemostasis research, including such landmark *in vivo* and *in vitro* studies as those of Martin (1894) and Mellanby (1909), who utilized respectively *Pseudechis porphyriacus* and *Notechis scutatus* venoms to study blood coagulation.

MECHANISMS OF VENOM ACTION

Following envenomation, a haemorrhagic diathesis could be the result of one or more of a variety of mechanisms (Table 17.1). Examples are:

a. Non-coagulant platelet-active substances, causing dysfunction, aggregation and/or thrombocytopenia and reviewed by

Brinkhous and Smith (1988) and Teng and Ma (1988). Further discussion is beyond the scope of this presentation. Many Australian snake venoms can aggregate platelets (Marshall and Herrmann, 1989), yet potent inhibitors of platelet aggregation can also be isolated from the same venoms.

b. Substances with a direct action on blood vessel walls which may interfere with endothelial cell integrity (haemorrhagins), as described by McKay *et al.* (1970), in experiments with *Vipera palestinae* venom. A number of other haemorrhagic proteases have been reported (Tu, 1988).

c. Anticoagulants, which may (a) possess activity of basic phospholipase A_2 or (b) directly inhibit serine proteinases of the coagulation cascade. Their actions have yet to be fully elucidated, but apparently they inhibit membrane-associated complexes which occur during coagulation as activators of factor X and prothrombin.

d. Activation of fibrinolysis, whether by direct release of tissue plasminogen activator from endothelial cells, the presence of a plasminogen activator or the activation of the intrinsic fibrinolytic system via kallikrein.

e. A consumptive coagulopathy caused by either a thrombin-like enzyme or an activator of the coagulation cascade.

f. The release into the circulation of tissue factor following tissue damage such as myonecrosis.

In addition to the above, proteinase inhibitors resembling aprotinin have been noted (e.g. *V. russelli*) and we have isolated inhibitors of this type from *Oxyuranus scutellatus* and *Pseudonaja textilis.*

It is widely considered that the haemostatically active components of venoms are not in themselves lethal, although a severe bleeding diathesis could cause fatal haemorrhage. Indeed, a number of thrombin-like enzymes derived from Crotalidae and Viperidae, which can induce defibrination, have been evaluated in the treatment of thrombosis. However, dangerous renal cortical necrosis has been recognized as a manifestation of disseminated intravascular coagulation following envenomation by various snakes and we have reported a patient with heavy *Pseudonaja textilis* envenomation in whom the clinical, laboratory, and pathologic features indicated that death was probably the direct result of the procoagulant. Subsequently we have studied

Table 17.1 Spectrum of haemostasis-active components in snake venoms

Class	*Comment*	*Examples*
Thrombin-like enzymes	Cleave fibrinopeptide A	*Agkistrodon rhodostoma*
		Bothrops atrox
	Cleave fibrinopeptide B	*Agkistrodon contortrix*
Factor X activator	Metalloproteinase	*Vipera russelli*
Prothrombin activator	Produce meizothrombin	*Echis carinatus*
(Metalloproteinase)		*Rhabdophis tigrinis*
Prothrombin activator	Factor X_a-like	*Notechis scutatus*
	Produce alpha-thrombin	*Tropidechis carinatus*
	Complete prothrombinase	*Pseudonaja textilis*
	Produce alpha-thrombin	*Oxyuranus scutellatus*
Protein C activator	Used for protein C assay	*Agkistrodon contortrix*
Haemorrhagin	Direct endothelial injury	*Crotalidae*
Inhibitors of activation of coagulation enzymes	Some are phospholipase A_2	*Naja* species
	Action upon membrane-associated enzyme complexes	*Pseudechis australis*
Inhibitor neutralization	Complex with antithrombin III etc.	*Agkistrodon acutus*
Fibrinogenase	Alpha chain degraded	*Naja* species
	(Metalloproteinase)	*Agkistrodon* species, etc.
	Beta Chain Degraded	*Crotalus atrox*
	(Serine proteinase)	*Trimeresurus* species
Plasminogen activator release	Direct endothelial action	Crotalidae
Aprotinin-like inhibitor	Plasmin inhibition	*Pseudonaja textilis*
		Oxyuranus scutellatus
Platelet aggregation		*Trimereserus okinavensis*
		Bothrops jararaca

purified preparations of prothrombin activators from *P. textilis* and from *Tropedichis carinatus*, showing that minute quantities (approximately 2 mg) are rapidly lethal to rats if administered intravenously, that lethality is prevented by preheparinization and that the induction of tolerance by exposure to initially small

(0.1 mg) then progressively increasing quantities of venom leads to a non-fatal coagulopathy with resistance to large quantities of procoagulant (100 mg+). Such data show a correlation between the clinical impact of the haemostatically active components of venoms with the quantity of venom administered and the rate and route of administration (whether it is injected so as to enter the circulation gradually, as it would from a tissue depot, or directly into a blood vessel as a bolus).

Clinically, we have invariably found that consumptive coagulopathy is present in patients significantly envenomated with *O. scutellatus*, *N. scutatus*, *T. carinatus*, *P. textilis* and, in one case, *O. microlepidotus* (sample provided by Dr Julian White). In such patients thrombocytopenia occurs but appears to be less extreme than expected by the severity of the coagulation disturbance. When checked, prolongation of bleeding time or abnormal platelet aggregation studies have shown the presence of platelet dysfunction, which could be caused either directly by an inhibitory venom component, by an acquired storage pool defect related to the consumptive coagulopathy, or by a combination of these. These patients have been studied by affinity chromatography of their plasma or serum using anti-fibrinogen serum bound to Sepharose 4B. Such studies have shown virtually total conversion of circulating fibrinogen into non-crosslinked and crosslinked fibrin degradation products, a result also confirmed by very high assays of D dimer using the monoclonal antibody DD-3B6/22 (Whitaker *et al.*, 1984). It can be deduced from these observations that thrombin is generated *in vivo* leading to the clotting of fibrinogen, the activation of factor XIII and stimulation of endogenous fibrinolysis. In spite of the profound coagulopathy, thrombosis cannot usually be demonstrated, probably reflecting the gradual induction of the defibrination accompanied by fibrinolysis (Whitaker *et al.*, 1968).

Thrombin-like enzymes are commonly found in crotaline snakes but have not been recorded in Australian elapid snake venoms. While their importance as defibrinating agents in treating thrombosis has not yet been firmly established, they remain of great theoretical interest. Whereas the majority appear to cleave fibrinopeptide A alone, an enzyme from *Ancistrodon contortrix* cleaves the B fibrinopeptide preferentially.

The proteinases described in venoms appear to fall into two broad groups, the metalloproteinases found especially in Colubridae and serine proteinases found mainly in the Elapidae, including the Australian varieties. The metalloproteinases include the factor X activator isolated from the venom of *V. russelli*, the prothrombin activators found in *Echis carinatus* and *Agkistrodon acutus* as well as proteinases which appear to destroy plasma inhibitors including antithrombin III, alpha$_2$ plasmin inhibitor and C1-INH. The serine proteinases include many thrombin-like enzymes, activators of kallikrein and complement, including the protein C activator from *Agkistrodon contortrix* and the activators of prothrombin and factor XIII, widely distributed in Australian snake venoms, which are discussed later.

Fibrinogenolysis can occur in response to fibrinogenases which act through degradation of the A-alpha chains of fibrinogen (Naja species) or degradation of the B-beta chains of fibrinogen (e.g. *Crotalus atrox*), or after epithelial release of plasminogen activator (described for *C. atrox* and *C. adamanteus*). The enzymes which degrade the A-alpha chain are all metalloproteinases and in addition to Naja species include various species of Agkistrodon, Trimersurus, Crotalus, Cerestes and some vipers. The enzymes degrading the B-beta chains are serine proteinases (e.g. *C. atrox* and *T. gramineus*).

Nonenzymic anticoagulants which bind plasma factors have been described with *A. acutus*. Enzymic anticoagulants are essentially the basic phospholipases A$_2$ common to many venoms. The presence of an inhibitor which becomes dissociated from procoagulant activity with venom dilution (Figure 17.1) is readily demonstrable in many, if not all, Australian snakes that have a procoagulant component (Masci, 1986). The venom of *Pseudechis australis* contains a large amount of anticoagulant activity (25–30% of total protein) and this has been partially purified, using in sequence, heparin Sepharose 4B chromatography and gel filtration with Sephacryl S-100. Two single chain components with phospholipase A$_2$ activity have been isolated and called inhibitor 1 and inhibitor 2 (Madaras *et al.*, 1983), having molecular weights of approximately 18 kD and 27 kD respectively on SDS-PAGE. These purified preparations prolong the prothrombin time, activated partial thromboplastin time, and platelet factor 3 time of citrated plasma but do not affect the thrombin clotting time. There is substantial, i.e. >50%, inhibition of the generation of factor X$_a$ by both the tissue factor pathway, and the intrinsic pathway, as well as the activation of prothrombin by X$_a$ or Taipan venom. They do not inhibit Russell's viper venom, activation of plasminogen by streptokinase, the activation of kallikrein, or the

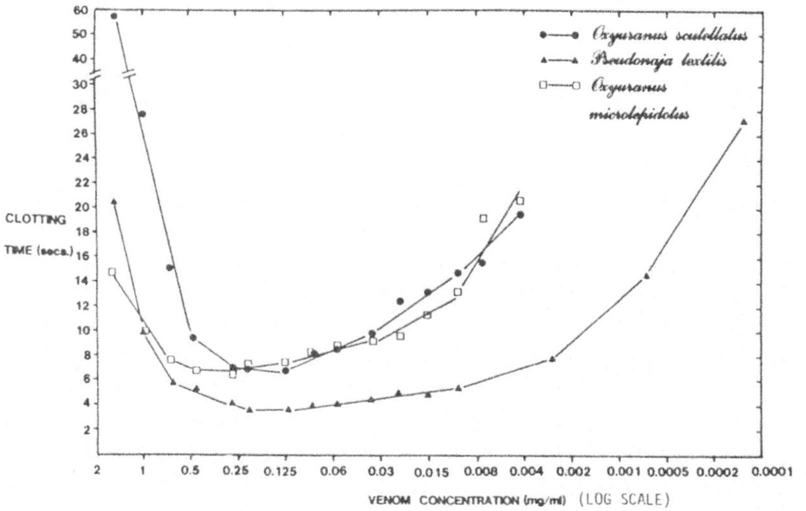

Figure 17.1 Effect of venom dilution upon the clotting of citrated plasma. Clotting times shorten initially with dilution of inhibitor, then lengthen again with dilution of prothrombin activator.

amidolytic activity of factor X_a on S 2222, of kallikrein on chromozym PK or of trypsin upon chromozym TRY. They are also potent inhibitors of platelet aggregation and of serotonin release, except for the first phase aggregation induced by ADP. *In vivo* they are anticoagulants, although higher doses of existing preparations may show some toxicity on intravenous injection, which might be caused by myotoxic or procoagulant contaminants.

There are three basic types of prothrombin activator. The metalloproteinase activators, such as in *E. carinatis* or *Rhabdophis tigrinis*, do not cleave the activation peptides from prothrombin and result in the formation of meizothrombin. The prothrombin activators of Australian snakes fall into two general groups. These both cleave prothrombin into alpha thrombin (Figure 17.2) and are capable of cleaving the N-terminal preactivation peptides including the GLA-domains, essentially as is caused physiologically by activated factor X. One type of activator, which requires the addition of calcium ions, phospolipid and activated factor V for optimal activity, is itself analogous to activated factor X. Such activators had been isolated from the venoms of *T. carinatus* and *N. scutatus*. The former has a molecular weight of approximately 54 kD by SDS-PAGE, consisting of chains of 18 and 37 kD

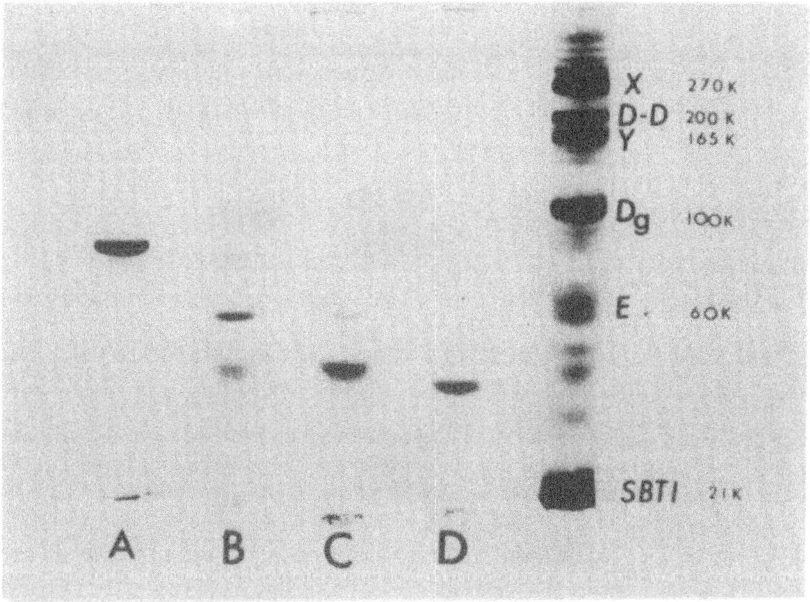

Figure 17.2 SDS-PAGE showing the cleavage of purified prothrombin by *P. textilis* venom (20 mg/ml) to produce alpha-thrombin. A 0.0 min; B 0.5 min; C 2.0 min and D 120.0 min. A marker gel is on the right.

respectively. This appears similar to the light and heavy chains of activated factor X, which contain respectively the GLA-domains and the active serine site. The venom activators of *O. scutellatus* and *P. textilis* are more complex. Whereas the prothrombin activator is a minor component of *O. scutellatus* venom, it appears to be the major protein component of *P. textilis* venom. These activators are complexes of molecular weight at least 200 kD which dissociate on SDS-PAGE to give a 64 kD molecule containing the active serine site and further dissociate on reduction into a two chain molecule (Figure 17.3). Detailed studies of the *O. scutellatus* venom procoagulant have been reported by Speijer *et al.* (1986) and Walker *et al.* (1980). The activator of *P. textilis* has been studied by Masci *et al.* (1988). These prothrombin activators all demonstrate amidolytic activity with the factor X*a* substrate, S 2222, and show an optimal calcium dependency of about 15 mM in citrated plasma. They all clot warfarin plasma indicating that they act against descarboxyprothrombin. Unlike *O. scutellatus, P. textilis* activator does not show phospholipid dependency

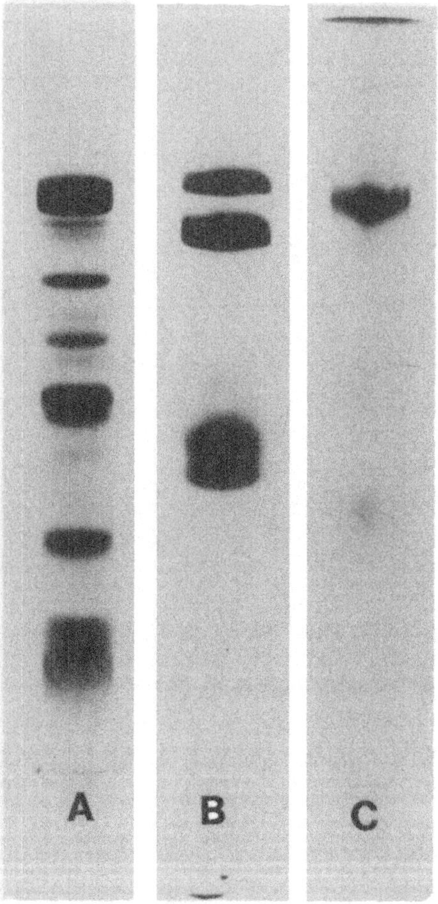

Figure 17.3 PAGE of *P. textilis* venom. A: Reconstituted dried venom, and B: purified prothrombin activator complex, both in the presence of SDS. In the absence of SDS (gel C), the activator appears as a single band on PAGE.

and is inhibited by benzamidine but not soy bean trypsin inhibitor. Antigenically, these two activators show near identity using both polyclonal (Figure 17.4) and monoclonal antibodies, and are more closely related to one another than to *O. microlepidotus*, contrary to other taxonomic data. We have prepared one monoclonal antibody which is specific for *P. textilis* activator alone.

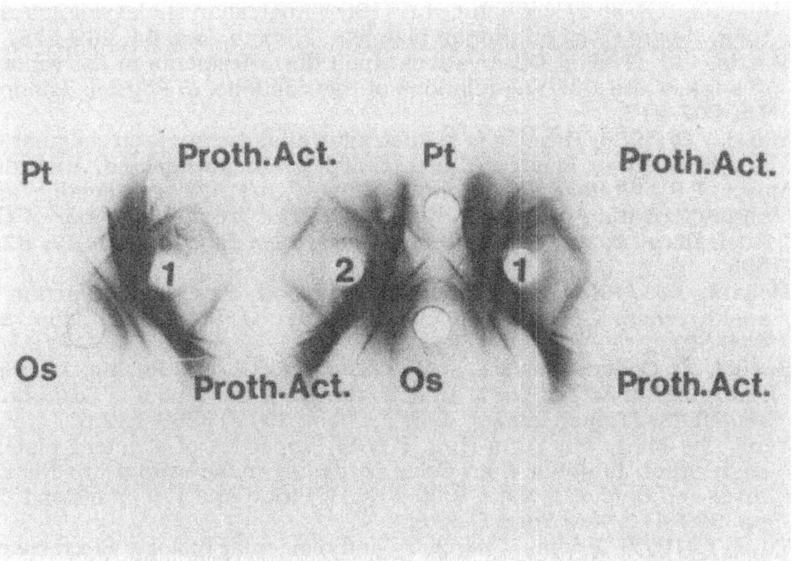

Figure 17.4 Ouchterlony plate of rabbit antiserum to (1) *P. textilis* venom and (2) *O. scutellatus* venom, diffused against the following antigens: Pt = *P. textilis* whole venom; Proth. Act. = purified prothrombin activator of *P. textilis*; Os = *O. scutellatus* whole venom. (Stained with coomassie blue).

CONCLUSIONS

After one century of study we still have much to do before understanding the biochemical complexities of the haemostatically-active components of Australian snake venoms. This brief overview does little justice to this complex and expanding subject.

REFERENCES

Brinkhous, K.M. and Smith, S. V. (1988) Platelet aggregating noncoagulant snake venom fractions: An overview. In: *Hemostasis and animal venoms*, edited by H. Pirkle, F.S. Markland Jr., pp. 363–376, New York: Dekker.

McKay, D.G., Moroz, C., DeVries, A., Csavossy, I. and Cruse, V. (1970) The action of hemorrhagin and phospholipase derived from Vipera palestinae venom on the microcirculation. *Lab. Invest.* **22**, 387–399.

Madaras, F., Smith, I.L. and Parkin, J.D. (1983) Purification and characterization of coagulation inhibitors from the King Brown venom. *Clin. Exp. Pharmacol. Physiol.* **10**, 490.

Marshall, L.R. and Herrmann, R.P. (1989) Australian snake venoms and their *in vitro* effect on human platelets. *Thromb. Res.* **54**, 269–275.

Martin, C.J. (1894a) Observations upon fibrin-ferments in the venoms of snakes and the time-relations of their actions. *J. Physiol.* (London) **15**, 207–215.

Masci, P.P. (1986) The effects of Australian snake venoms on coagulation and fibrinolysis. Masters Thesis, University of Queensland, Australia.

Masci, P.P., Whitaker, A.N. and DeJersey, J. (1988) Purification and characterization of a prothrombin activator from the venom of the Australian brown snake, *Pseudonaja textilis*. *Biochem. Int.* **17**, 825–835.

Mellanby, J. (1909) The coagulation of blood. Part 11. The action of snake venoms, peptone and leech extract. *J. Physiol.* (London) **38**, 441–503.

Speijer, H., Govers-Riemslag, J.W.P., Zwaal R.F.A. and Rosing, J. (1986) Prothrombin activation by an activator from the venom of *Oxyuranus scutellatus* (Taipan snake). *J. Biol. Chem.* **261**, 13258–13267.

Teng Che-Ming, Ma Yunn-Hwa. (1988) Properties of a potent platelet aggregation inhibitor from *Echis carinatus* snake venom. In: *Hemostasis and animal venoms*, edited by H. Pirkle and F.S. Markland Jr., pp. 399–410, New York: Dekker.

Tu, A.T. (1977) Venoms, Chemistry and molecular biology. *Blood coagulation* **21**, 329–360.

Walker, P.J., Owen, W.G., and Esmon, C.T. (1980) Characterization of the prothrombin activator from the venom of *Oxyuranus scutellatus* (Taipan venom). *Biochem.* **19**, 1020–1023.

Whitaker, A.N., Elms, M.J., Masci, P.P., Bundesen, P.G., Rylatt, D.B., Webber, A.J. and Vunce, I.H. (1984) Measurement of crosslinked fibrin derivatives in plasma: An immunoassay using monoclonal antibodies. *J. Clin. Pathol.* **37**, 882–887.

Whitaker, A.N., McKay, D.G. and Csavossy, I. (1969) Disseminated intravascular coagulation. *Am. J. Pathol.* **56**, 153–176.

18 A TOXIC PROTEIN FROM THE AIDS VIRUS

ROSS I. BRINKWORTH, ANNELIESE F. PALMER, STEPHEN L. RAMSDALE and PETER R. ANDREWS

INTRODUCTION

The initial event in the binding of the AIDS virus, HIV, to "helper" T-lymphocytes, is the binding of the HIV envelope glycoprotein, gp120, to the helper T-cell-specific antigen, CD4 (McDougal *et al.*, 1986). The envelope glycoprotein, gp120, and the transmembrane glycoprotein, gp41, are produced as a precursor, gp160, which is subsequently cleaved by a cellular protease. The gp120 from the HIV-1 isolate, WMJ1, has 481 amino acids and 20 potential asparagine-linked glycosylation sites, NXS or NXT (underlined in the sequence below) (Starcich *et al.*, 1986).

POSSIBLE TARGET SITES FOR GP120

The expression of gp120 on the surface of HIV-infected cells is responsible for cytotoxic effects which can destroy T-lymphocytes in a mechanism separate from viral replication. A direct antibody-dependent cellular toxicity may arise from the similarity between expressed gp120 on infected cells and MHC class II proteins (HLA-DR, HLA-DQ etc.) carrying bound foreign antigen (Evans *et al.*, 1989, Tyler *et al.*, 1989, Weinhold *et al.*, 1989). When the amino acid sequences of various class II MHC alleles and the gp120 sequences from different isolates were compared, the best match was achieved when HLA-DQ2 (Boss and Strominger, 1984) was matched with the gp120 from the WMJ1 strain of HIV-1 (Starcich *et al.*, 1986). The sequence of gp120 [WMJ1] from

Cys415 to Ser451 shows partial homology with the sequence Cys15 to Ser63 of HLA-DQb1 (Brinkworth, 1989):

HLA-DQb1, 15-61:

CYFTNGTERVRLVSRSIYNREEIVRFDSDVGEFRAVTLLGLPAAEYWNS

gp120 [WMJ1], 415–451:

CSSNITGLLLTRDGGNSSSREEIFRPG_G_GNMR__________DNWRS

This mimicry is also responsible for the binding of gp120 to CD4, since class II MHC proteins are the natural ligands of CD4. A major cause of cell death in HIV infection is the formation of syncytia or short-lived, giant, multi-nucleated cells produced through an interaction between gp120 expressed on the surface of infected cells and CD4 on the surface of non-infected cells. Helpter T-lymphocytes which have been specifically produced to combat the AIDS virus may express T-cell receptor/CD4 complexes which are targeted to antigen-presenting cells carrying class II MHC proteins with bound gp120 epitopes. Such T-lymphocytes may therefore recognize infected, gp120-presenting cells in error and thus be selectively eliminated.

As well as the CD4 binding function, gp120 is cytotoxic in its own right (Barnes, 1987). AIDS-related dementia is due to a neurotoxicity which may or may not be associated with CD4 antigen. AIDS-related dementia is in the long term potentially more serious for society than the immunodeficiency and may not necessarily respond to the same treatments as the immunodeficiency. gp120 is toxic to animal nerve cells *in vitro* and has an unusual dose–response curve in that it is most potent at intermediate concentrations (Barnes, 1987). A recent study has shown that CD4 is not the receptor for HIV-1 infection of several neural cell lines (Harouse *et al.*, 1989). Instead, the neurotoxic effects of gp120 have been attributed to either an antagonism of the neurotrophic actions of Vasoactive Intestinal Peptide (VIP) on neuroglial cells or an interference with the growth-promoting effects of Neuroleukin.

Vasoactive Intestinal Peptide is a potent bioactive peptide of 28 amino acids which has been shown to have activity in the gut, the cardiovascular system, the central nervous system, the respiratory system and the immune system (Fahrenkrug and Emson, 1982). It has the amino acid sequence:

HSDAVFTDNYTRLRKQMAVKKYLNSILN-NH_2

The transduction mechanism of VIP is the stimulation of production of intracellular 5',6'-cyclic AMP resulting in smooth muscle relaxation and neurostimulation. VIP coexists with acetylcholine in neurons innervating salivary glands. In the brain, VIP is found in highest concentrations in the cerebral cortex, hippocampus, amygdala, the terminal nuclei of the interstitial striae, the suprachiasmatic nucleus, the superior colliculus, the olfactory bulb, the solitary nucleus, the ventral part of the mesencephalic grey, and the corpus striatum. In the cerebral cortex, it is located in uniformly distributed bipolar neurons running perpendicular to the cortical surface (Nieuwenhuys, 1985). Brain VIP receptors have a similar distribution.

There is considerable circumstantial evidence that VIP may have a role in the development of HIV-induced pathology. Firstly, VIP receptors have a very similar distribution across regions of the brain to that reported for CD4 (Hill *et al.*, 1986; Maddon *et al.*, 1986; Pert *et al.*, 1986). In addition, VIP has a neurotrophic action in protecting neurons from both tetrodotoxin-mediated (Brenneman *et al.*, 1987) and gp120-mediated (Brenneman *et al.*, 1988a) cell death and like CD4, VIP receptors are marker antigens for helper T-lymphocytes (Fantini *et al.*, 1985). VIP is also immunosuppressive, inhibiting lymphocyte migration, decreasing the mitogenic response of T-cells (Ottaway, 1987), and decreasing the production of interleukin-2 by stimulated lymphocytes (Morley *et al.*, 1987). An intriguing possibility is that the trypsin-like extracellular enzyme which inactivates VIP on human lymphocytes (Goetzl *et al.*, 1989) is the same enzyme which converts gp160 to gp120 and gp41.

VIP cleavage point:	YTRL**R**/**K**QMAV
gp160 cleavage point:	VQRE**K**/**R**AVGA

However, there are a number of facts which do not support VIP being directly involved in AIDS. The membrane transduction system for VIP receptors is the stimulation of cAMP production unlike CD4 which is linked to the calcium/phosphoinositol pathway. VIP is immunosuppressive whereas gp120 by itself is actually a T-cell activator (Kornfeld *et al.*, 1988).

Whilst there is some evidence for VIP being involved in AIDS, there is considerable doubt that Peptide T, the octapeptide derived from the gp120 from the ARV2 isolate of HIV, is the active component. Peptide T is an octapeptide found in the amino acid sequence of the envelope glycoprotein (gp120) of the ARV-2 strain of HIV-1 (Pert *et al.*, 1986). Peptide "T" was named because of its high concentration of Threonine (T) residues, its sequence being ASTTTNYT.

There is substantial homology of Peptide T with VIP:

VIP:	HSD**A**VF**T**D**NYT**RLRKQMAVKKYLNSILN-NH_2
Peptide T:	**ASTTTNYT**

Pert and associates at NIMH, Maryland have proposed that Peptide T forms part of the binding site for CD4 and that TTNYT is the "active" portion. They further proposed that like VIP, Peptide T can protect against gp120-mediated neuronal cell death (Brenneman *et al.*, 1988b). Picone and coworkers have demonstrated, by the use of 500 MHz proton NMR, that a prominent conformational feature of Peptide T is an N-terminal β-bend incorporating the last four residues TNYT (Picone *et al.*, 1988, Motta *et al.*, 1989). This has been confirmed by work in our laboratory.

One of the main objections to the theory that Peptide T is involved in CD4 binding is that Peptide T is in a hypervariable section of gp120. Consider the equivalent region in a number of HIV-1 isolates (Starcich *et al.*, 1986):

gp120 [ARV2] 188–212	NLDVVPIDNASTTTNYTNYRLIHC
gp120 [WMJ1]	KLDVVPI KSNNDSST YRLIHC
gp120 [BH10]	KLDIIPIDND TTS YTLTSC
gp120 [BR]	KLDIVPIDNDK TN TRYRLISC
gp120 [Z-6]	RLDIVPIDNDNSSTNSTNYRLINC

In addition, the N-terminal half of gp120 is not believed to be involved in CD4 binding. With the exception of the initial findings (Pert *et al.*, 1986) most of the positive results using the Peptide T have arisen from experiments using a lymphocyte chemotaxis assay (Ruff *et al.*, 1987; Sacerdote *et al.*, 1987; Pert *et al.*, 1988). The relevance of this to HIV is not at all clear. Positive anti-HIV results from experiments on Peptide T have not been able to be reproduced by other groups (Sodroski *et al.*, 1987). It has also been shown that Peptide T does not interact with VIP receptors (Nguyen, 1988, Yuwiler and Wetterberg, 1989).

Thymosin α1 has the sequence Acetyl-SDAAVDTSSEITTK-DLKEKKEVVEEAEN. It has been shown that circulating levels of Thymosin α1 are elevated in patients with AIDS. Thymosin α1 stimulates the secretion of several lymphokines, including MIF, α- and γ-interferons, lymphotoxin and T-cell growth factor. It can amplify T-cell immunity by stimulating helper cell function (Nguyen and Scheving, 1987). It has been proposed that Peptide T could act as an analogue of part of Thymosin α1 [ASTTTN-YT vs AVDTSSEIT] (Nguyen and Scheving, 1987). Thymosin α1 may behave as a partial agonist of VIP and there is evidence for the specific binding of Thymosin α1 to VIP receptors in rat monocyte and liver plasma membranes (Calvo *et al.*, 1986).

Other immunoactive substances have been implicated in the activity of gp120. Interleukin 2 is a lymphocyte growth factor produced by T-cells that is essential for a normal immune response. Its chief effect on lymphoid cells is proliferation and differentiation. IL-2 has 133 amino acid residues. Part of gp41 is homologous with IL-2 (Reiher *et al.*, 1986):

IL-2, 34–39:	LEHLLL
gp41[ARV-2], 850–855:	LERLLL

Neuroleukin is secreted by activated T lymphocytes and induces immunoglobulin secretion by human peripheral blood monocytes. It is a neurotrophic factor for spinal and sensory neurons. An homology with gp120 has been demonstrated (Lee *et al.*, 1987):

Neuroleukin (NLK) 403–407:

PCDFLIP**VQ**TQ**H**P I**R**KGLHHKI**LL**ANF**LA**QTEALM**R**GKSTEEAR**K**

gp120 [WMJ1] 238–282:

PCTNVST**VQ**CT**H**GI**R**PVVSTQL**LL**NGS**LA**EEEIVI**R**SENFTDNA**K**

The significance of this finding has been brought into question by the discovery that the proposed sequence of NLK is identical to that of phosphohexose isomerase (Chaput *et al.*, 1988, Faik *et al.*, 1988).

CONCLUSION

It is to be expected that the question of the involvement of receptors other than CD4 in the etiology of HIV infection will be clarified in the not too distant future. Aside from that, the exact role of the VIP-Thymosin α1 axis in the immune system and in the interaction between the immune system and the nervous system needs further elucidation.

REFERENCES

Barnes, D.M. (1987) Solo actions of AIDS virus coat. *Science* **237**, 971–973.

Boss, J.M. and Strominger, J.L. (1984) Cloning and sequence analysis of the human histocompatibility complex gene DC-3b. *Proc. Natl. Acad. Sci. USA* **81**, 5199–5203.

Brenneman, D.E., Buzy, J.M., Ruff, M.R. and Pert, C.B. (1988b) Peptide T sequences prevent neuronal cell death produced by the envelope protein (gp120) of the human immunodeficiency virus. *Drug. Devel. Res.* **15**, 361–369.

Brenneman, D.E., Neale, E.A., Foster, G.A., D'Autremont, S.W. and Westbrook, G.L. (1987) Nonneuronal cells mediate the neurotrophic action of vasoactive intestinal peptide. *J. Cell. Biol.* **104**, 1603–1610.

Brenneman, D.E., Westbrook, G.L., Fitzgerald, S.P., Ennist, D.L., Elkins, K.L., Ruff, M.R. and Pert, C.B. (1988a) Neuronal cell killing by the envelope protein of HIV and its prevention by vasoactive intestinal peptide. *Nature* **335**, 639–642.

Brinkworth, R.I. (1989) The envelope glycoprotein of HIV-1 may have incorporated the CD4 binding site from HLA-DQb1. *Life Sciences* (in press).

Calvo, J.R., Goberna, R. and Guerrero, J.M. (1986) Interaction of thymic peptide Thymosin a1 with vasoactive intestinal peptide (VIP) receptors. *Bioscience Reports* **6**, 727–733.

Chaput, M., Claes, V., Portetelle, D., Cludts, A., Cravador, A., Burny, A., Gras, H. and Tartar, A. (1988) The neurotrophic factor neuroleukin is 90% homologous with phosphohexose isomerase. *Nature* **332**, 454–455.

Evans, L.A., Thomson-Honnebier, G., Steimer, K., Paoletti, E., Perkus, M.E., Hollander, H. and Levy, J.A. (1989) Antibody-dependent cellular toxicity is directed against both the gp120 and gp41 envelope proteins of HIV. *AIDS* **3**, 273–276.

Fahrenkrug, J. and Emson, P.C. (1982) Vasoactive intestinal polypeptide: functional aspects. *Br. Med. Bull*. **38**, 265–270.

Faik, P., Walker, J.I.H., Redmill, A.A.M. and Morgan, M.J. (1988) Mouse glucose-6-isomerase and neuroleukin have identical 3' sequences. *Nature* **332**, 455–457.

Fantini, J., Tirard, A., El Battari, A., Luis, J., Muller, J.-M., Pichon, J. and Marvaldi, J. (1985) Le récepteur du VIP sur le Lymphocyte T: un nouveau marquer de différenciation? *Gastroenterol. Clin. Biol.* **9**, 929–935.

Goetzl, E.J., Kodama, K.T., Turck, C.W., Schiogolev, S.A. and Sreedharan, S.P. (1989) Unique pattern of vasoactive intestinal peptide by human lymphocytes. *Immunology* **66**, 554–558.

Harouse, J.M., Kunsch, C., Hartle, H.T., Laughlin, M.A., Hoxie, J.A., Wigdahl, B. and Gonzalez-Scarano, F. (1989) CD4-independent infection of human neural cells by human immunodeficiency virus type 1. *J. Virol.* **63**, 2527–2533.

Hill, J.M., Farrar, W.L. and Pert, C.B. (1986) Localisation of the T4 antigen/AIDS virus receptor in monkey and rat brain: prominence in cortical regions. *Psychopharmacology Bulletin* **22**, 689–694.

Kornfeld, H., Cruikshank, W.W., Pyle, S.W., Berman, J.S. and Center, D.M. (1988) Lymphocyte activation by HIV-1 envelope glycoprotein. *Nature* **335**, 445–448.

Lee, M.R., Ho, D.D. and Gurney, M.E. (1987) Functional interaction and partial homology between human immunodeficiency virus and neuroleukin. *Science* **237**, 1047–1051.

McDougal, J.S., Kennedy, M.S., Sligh, J.M., Cort, S.P., Mawle, A. and Nicholson, J.K.A. (1986) Binding of HTLV-II/LAV to T4+ T cells by a complex of the 110K viral protein and the T4 molecule. *Science* **231**, 382–385.

Maddon, P.J., Dalgleish, A.G., McDougal, J.S., Clapham, P.R., Weiss, R.A. and Axel, R. (1986) The T4 gene encodes the AIDS virus receptor and is expressed in the immune system and brain. *Cell* **47**, 333–348.

Morley, J.E., Kay, N.E., Solomon, G.F. and Plotnikoff, N.P. (1987) Neuropeptides: conductors of the immune orchestra. *Life Sciences* **41**, 527–544.

Motta, A., Picone, D., Temussi, P.A., Marastoni, M. and Tomatis, R. (1989) Conformational analysis of Peptide T and of its C-pentapeptide fragment. *Biopolymers* **28**, 479–486.

Nguyen, T.D. (1988) Peptide T from human immunodeficiency virus envelope does not interact with hepatic, intestinal and colonic vasoactive intestinal peptide (VIP) receptors. *Peptides* **9**, 425–428.

Nguyen, T.D. and Scheving, L.A. (1987) Thymosin a1: amino acid homology with peptide T from the human immunodeficiency virus envelope. *Biochem. Biophys. Res. Commun.* **145**, 884–887.

Nieuwenhuys, R. (1985) *Chemoarchitecture of the brain*, pp. 57–60. Berlin: Springer-Verlag.

Ottaway, C.A. (1987) Selective effects of vasoactive intestinal peptide on the mitogenic response of murine T cells. *Immunology* **62**, 291–297.

Pert, C.B., Hill, J.M., Ruff, M.R., Berman, R.M., Robey, W.G., Arthur, L.O., Ruscetti, F.W. and Farrar, W.L. (1986) Octapeptides deduced from the neuropeptide receptor-like pattern of antigen T4 in brain potently inhibit human immunodeficiency virus receptor binding and T-cell infectivity. *Proc. Natl. Acad. Sci. USA* **83**, 9254–9258.

Pert, C.B., Smith, C.C., Ruff, M.R. and Hill, J.M. (1988) AIDS and its dementia as a neuropeptide disorder: role of VIP receptor blockade by human immunodeficiency virus envelope. *Ann. Neurol.* **23** (suppl.), S71–S73.

Picone, D., Temussi, P.A., Marastoni, M., Tomatis, R. and Motta, A. (1988) A 500 MHz study of peptide T in a DMSO solution. *FEBS Letts.* **231**, 159–163.

Reiher, W.E., Blalock, J.E. and Brunck, T.K. (1986) Sequence homology between acquired immunodeficiency syndrome virus envelope protein and interleukin 2. *Proc. Natl. Acad. Sci. USA* **83**, 9188–9192.

Ruff, M.R., Martin, B.M., Ginns, E.I., Farrar, W.L. and Pert, C.B. (1987) CD4 receptor binding peptides that block HIV infectivity cause human monocyte chemotaxis. *FEBS Letts.* **211**, 17–22.

Sacerdote, P., Ruff, M.R. and Pert, C.B. (1987) Vasoactive intestinal polypeptide 1–12: a ligand for the CD4 (T4)/human immunodeficiency virus receptor. *J. Neuroscience Research* **18**, 102–107.

Sodroski, J., Kowalski, M., Dorman, T., Basiripour, L., Rosen, C. and Haseltine, W. (1987) HIV envelope -CD4 interation not inhibited by synthetic octapeptides. *Lancet* June 20, 1428–1429.

Starcich, B.R., Hahn, B.H., Shaw, G.M., McNeely, P.D., Modrow, S., Wolf, H., Parks, E.S., Parks, W.P., Josephs, S.F., Gallo, R.C. and Wongstaal, F. (1986) 5 Identification and characterization of conserved and variable regions in the envelope gene of HTLV/LAV, the retrovirus of AIDS. *Cell* **45**, 637–648.

Tyler, D.S., Nastala, C.T., Stanley, S.D., Matthews, T.J., Lyerly, H.K., Bolognesi, D.P. and Weinhold, K.J. (1989) gp120 specific cellular toxicity in HIV-1 seropositive individuals. *J. Immunol.* **142**, 1177–1182.

Yuwiler, A. and Wetterberg, L. (1989) Peptide T does not affect induction of pineal N-acetyltransferase by vasoactive intestinal peptide. *Regulatory Peptides* **25**, 69–73.

Weinhold, K.J., Lyerly, H.K., Stanley, S.D., Austin, A.A., Matthews, T.J. and Bolognesi, D.P. (1989) HIV-1 gp120-mediated immune suppression and lymphocyte destruction in the absence of viral infection. *J. Immunol.* **142**, 3091–3097.

19 TOXICITY TESTING

GEOFFREY NEWMAN-MARTIN

INTRODUCTION

In Australia the assessment of agricultural and veterinary chemicals and pesticides is carried out under the federal *Agricultural and Veterinary Chemicals Act*, through the Australian Agricultural and Veterinary Chemicals Council (AAVCC). Toxicological assessments on these chemicals are carried out by expert committees of the National Health and Medical Research Council (NH & MRC). NH & MRC's recommendations are passed through AAVCC to the States and Territories, which incorporate them into legislation. These recommendations are in the form of poisons schedules, which control availability of these chemicals as well as aspects of labelling such as safety directions, and maximum residue limits (MRLs), which control maximum concentrations of pesticides or other substances allowed in various food commodities.

Toxicity studies required by NH & MRC are: toxicokinetic studies, acute toxicity testing, sub-chronic toxicity studies, chronic toxicity studies, oncogenicity (carcinogenicity) studies, reproduction and developmental toxicity (including teratology), and genotoxicity (mutagenicity) studies. Other studies such as neurotoxicity may also be required.

The bulk of information required for assessment is in the form of submissions from industry which contain detailed toxicology data and results of pesticide residue trials. NH & MRC does not itself conduct toxicology testing. Other information is obtained from published scientific literature, government agencies of other countries (e.g. US Environmental Protection Agency), international organizations (such as WHO/FAO Joint Meeting on Pesticide Residues), other government departments and consumer organizations.

TOXICOKINETIC STUDIES

Toxicokinetic studies are studies on absorption, distribution, metabolism and excretion of the chemical in the body of the experimental animal. Such studies assist in the design of animal toxicity studies and in the selection of the most appropriate species for further toxicity studies, and assist in extrapolating toxicity in animals to potential effects in humans. The species chosen are those used for toxicity studies; exposure routes should be those applicable to human exposure, although often the oral route is used for convenience.

ACUTE TOXICITY TESTING

Acute toxicity studies involve the administration of a single dose or multiple doses of the chemical, given within 24 hours, and the observation of adverse effects occurring from the time of administration up to 14 days post-exposure. Administration is carried out by various routes such as oral, dermal and inhalation. The median lethal dose or LD_{50} test is not required, but the highest dose should produce deaths. It is desirable that two species be used. Acute toxicity testing provides information on health hazards likely to arise from a short-term exposure by the relevant route. It is a useful rapid indication of the potential hazards of acute exposure; in terms of human exposure this represents an abnormal or accidental situation. The objectives of acute toxicity studies include the estimation of relative toxicity, and the identification of major mechanisms of toxicity and of the most hazardous routes of exposure.

Dermal and ocular irritation and corrosion tests are also carried out. Irritation is the production of reversible inflammatory changes; corrosion is the production of irreversible tissue damage following the application of a test substance. Skin sensitization studies are conducted in order to determine the potential of a chemical to cause allergic contact dermatitis. Acute toxicity studies of this type provide information which is useful in developing safety directions for product labelling.

SUB-CHRONIC TOXICITY TESTING

Sub-chronic toxicity studies examine the adverse effects occurring as a result of the repeated daily administration of a chemical to experimental animals by various routes of exposure. The studies are usually conducted over a 3-month period on two species (usually rat and dog), using three dose levels, which preferably include one toxic level. In sub-chronic toxicity studies, extensive monitoring of the test animals is required. The objectives of sub-chronic toxicity studies are to determine the potential hazard with repeated exposure, identify target organs, and to suggest dose-levels for chronic studies.

CHRONIC TOXICITY TESTING

Chronic toxicity studies determine the potential hazard of long-term low-level exposure. Such tests are carried out to ascertain the maximum dosage level which produces no discernible ill-effects when administered over the major portion of the life-span of the experimental animal, and to reveal effects not predictable from short-term tests. Two species are required, often rat and dog, with three dose-levels, the highest toxic and the lowest non-toxic. Extensive monitoring of experimental animals is required including full clinical chemistry, urinalysis and hematology during the study.

CARCINOGENICITY TESTING

Chronic toxicity studies may be combined with oncogenicity studies. At post mortem full gross and microscopic pathology examinations are conducted.

The objective of a carcinogenicity study is to observe test animals for a major portion of their life-span for the development of cancer after exposure to the test compound. Tests in two species are required with adequate numbers of animals for statistics. Rats and mice are used, because of their relatively short life-span, the limited cost of maintenance, their widespread use in pharmacological and toxicological studies, their susceptibility to tumor induction, and the availability of strains whose characteristics are well known. Test chemicals are given at the maximum tolerated dose, i.e. that dose which can be given in a chronic study and will not shorten the animals' life-span from effects other than carcinogenicity. However, doses should be selected so that some toxicity occurs. The experimental animals need to be extensively monitored for tumors.

TESTING FOR REPRODUCTIVE TOXICITY

The term reproduction and developmental toxicity covers the area of reproduction, fertility and teratogenicity. The experimental design of reproduction studies involves a feeding study on one species, usually the rat, in a multi-generation study, that is, for two or three generations, for two litters each. Information about effects on all aspects of reproduction in both sexes of the test animals, including sexual behavior, gonadal function, fertility, fecundity, parturition, lactation, pre- and post-natal growth, and development and maturation of the offspring is provided.

Developmental studies are required in two species, and give information on embryo lethality, teratogenicity and altered growth. Teratogenicity is the property of a chemical which causes permanent structural or functional abnormalities when administered during embryonic development. Determination of the teratogenic potential provides information on the potential hazard to humans, in terms of the unborn, which may arise from exposure of the mother to the chemical during pregnancy. Teratogenicity studies involve dosing (usually oral) during organogenesis, for two species, usually the rat and the rabbit. Three dose levels are used, the highest being toxic to the mother. The foetuses are examined for skeletal and visceral abnormalities.

MUTAGENICITY TESTING

Genotoxicity (mutagenicity) tests are used to determine the potential of a compound to cause genetic damage. They may be predictive of potential carcinogenicity for genotoxic carcinogens. Various tests are used which can detect gene point mutations in microbes, chromosomal damage in mammalian cells in culture, gene mutations in mammalian cells in culture, or chromosome damage in whole animals using either the mouse micronucleus test or metaphase analysis of bone marrow or other proliferative tissue.

Other studies may include special studies on neurotoxicity; the toxicity of metabolites and impurities; the toxicity of mixtures;

and human data such as poisonings, suicides, and occupational exposure studies and epidemiology from exposed workers.

CONCLUSION

From evaluation of the various toxicity studies one can calculate an Acceptable Daily Intake (ADI) of a pesticide residue, which is the daily intake which, during an entire lifetime, appears to be without appreciable risk to the health of the consumer on the basis of all facts known at the time. The ADI is based on a "no-effect level" in animals together with a safety factor, ranging from 10 when human data are available to 2000 when definitive chronic studies are lacking.

Chemical hazard is dependent on the intrinsic toxicity of the chemical and the potential for human exposure. The results of toxicity studies can be used by regulators to control public exposure to chemicals. This is done by controlling availability through poisons scheduling and encouraging correct handling through labelling. Exposure of the general public is controlled by standards for residues of agricultural and veterinary chemicals in foods, and by adherence to good agricultural practices. Exposure of workers can be limited by good occupational hygiene, use of protective clothing where appropriate, and adherence to safety directions.

20 THE EFFECTS OF CHLOROPROPANES ON SPERMATOZOA

MICHAEL DOBBIE AND ALAN JONES

INTRODUCTION

Chlorinated derivatives of propane, such as epi-chlorohydrin and the nematocide 1,2-dibromo-3-chloropropane, are beneficial to man but are inherently toxic. A common oxidative metabolite of these compounds is (*R*,*S*)-α-chlorohydrin (3-chloro-1,2-propan-diol) which is further oxidized to (*R*,*S*)-3-chlorolactate. The observed diuresis accompanied by glucosuria, and the excretion of oxalate in rats following sub-lethal doses of (*R*,*S*)-α-chlorohydrin, is attributed to (*R*)-3-chlorolactate or a further oxidative metabolite (Porter and Jones, 1982). Concurrent studies have shown that chloropropanes and hydroxychloropropanes have antifertility activity in the male, a response attributed solely to the (*S*)-isomers (Jones *et al.*, 1979). (*S*)-α-Chlorohydrin is oxidized within mature spermatozoa to (*S*)-3-chlorolactaldehyde which specifically inhibits triosephosphate isomerase and glyceraldehyde-3-phosphate dehydrogenase (Cooney and Jones, 1988). Chemical and biochemical oxidations *in vitro* provide circumstantial evidence that 3-chloropyruvate is an oxidative metabolite of (*R*)-3-chlorolactate *in vivo* (Dobbie *et al.*, 1988). We have investigated whether the acidic metabolites, (*R*)- and (*S*)-3-chlorolactate and 3-chloropyruvate, exhibit a toxic effect on the metabolism of mature spermatozoa.

RESULTS

(*R,S*)-3-Chlorolactate and 3-chloropyruvate have now been shown to inhibit the production of CO_2 from fructose by rat, guinea pig and boar spermatozoa. We have developed a model system for detailed investigation of these inhibitors using boar spermatozoa which are readily available from abattoir material (Jones and Chantrill, 1989).

Both (*R*)- and (*S*)-3-chlorolactate (5 mM each) as well as 3-chloropyruvate (1 mM) inhibited the production of CO_2 from fructose (30%, 21% and 93% inhibition after 1 h, respectively) while only the (*R*)-isomer and 3-chloropyruvate have an inhibitory effect on the oxidation of lactate to CO_2 (47% and 37% inhibition after 1 h, respectively). The inhibitory actions of (*R*)-3-chlorolactate and 3-chloropyruvate on the oxidation of various radiolabeled substrates were further investigated.

Both (*R*)-3-chlorolactate (5 mM) and 3-chloropyruvate (1 mM) immediately inhibited the oxidation of fructose and fructose-1,6-bisphosphate to CO_2, with 3-chloropyruvate being the more potent inhibitor, particularly in the case of fructose oxidation (Figure 20.1A). Transient inhibition of glycerol oxidation was evident with both acids (Figure 20.1B) while (*R*)-3-chlorolactate failed to inhibit glycerol-3-phosphate oxidation. Both compounds also inhibited the oxidation of lactate to CO_2, with the effect due to 3-chloropyruvate exhibiting a 5 min lag phase (Figure 20.1C). Pre-incubation of cells with either inhibitor before the addition of radiolabeled substrate resulted in the inhibition of fructose oxidation by (*R*)-3-chlorolactate becoming transient (Figure 20.2A) and the inhibitory effects of 3-chloropyruvate on the oxidation of fructose-1,6-bisphosphate, lactate (Figure 20.2C), and in particular glycerol (Figure 20.2B) and glycerol-3-phosphate, becoming enhanced.

Although both acids significantly inhibited CO_2 production from a variety of substrates neither has an effect on the energy charge of the cells as calculated from adenine nucleotide concentrations.

Stereochemical considerations suggest that (*R*)-3-chlorolactate could inhibit the activity of lactate dehydrogenase or malate dehydrogenase and that 3-chloropyruvate could act on lactate dehydrogenase, the pyruvate dehydrogenase complex or citrate synthase. Assays of intermediates following 1 h incubation with either (R)-3-chlorolactate (5 mM) or 3-chloropyruvate (1 mM) did not alter the concentrations of lactate, pyruvate, acetyl-CoA, oxaloacetate, or citrate but that of malate increased in the presence of (*R*)-3-chlorolactate and decreased in the presence of 3-chloropyruvate when lactate was offered as a substrate (Figure 20.3). Preliminary experiments indicate that (*R*)-3-chlorolactate and in particular 3-chloropyruvate may be inhibitors of lactate dehydrogenase purified from whole boar spermatozoa.

DISCUSSION

In mature boar spermatozoa both (*R*)-3-chlorolactate and 3-chloropyruvate inhibit the oxidation of several substrates to CO_2. A notable exception is the lack of effect of (*R*)-3-chlorolactate on glycerol-3-phosphate oxidation even though there is a transient inhibition of glycerol oxidation. There appears to be at least two sites of action by 3-chloropyruvate— at least one in the glycolytic pathway and another intra-mitochondrially since the inhibition of lactate oxidation is less than that of fructose. Also of note is the greatly enhanced inhibition of glycerol and glycerol-3-phosphate by 3-chloropyruvate following pre-incubation. Is 3-chloropyruvate being metabolized to a more potent inhibitor under these conditions or does pre-incubation simply allow 3-chloropyruvate uncompetitive access to the site of inhibition?

Furthermore, it appears that the cells are able to maintain a high energy state in the absence of CO_2 production. It is postulated that though the flux of the citric acid cycle is decreased, the generation of reducing equivalents (e.g. NADH) by the action of, for example lactate dehydrogenase, could maintain ATP levels and thus cellular viability.

CONCLUSION

Both of the acidic metabolites of chloropropanes which have been investigated are inhibitors of the oxidation of various substrates by mature boar spermatozoa. It is likely that (*R*)-3-chlorolactate and 3-chloropyruvate act at different and multiple, as yet unknown, sites in spermatozoal metabolism.

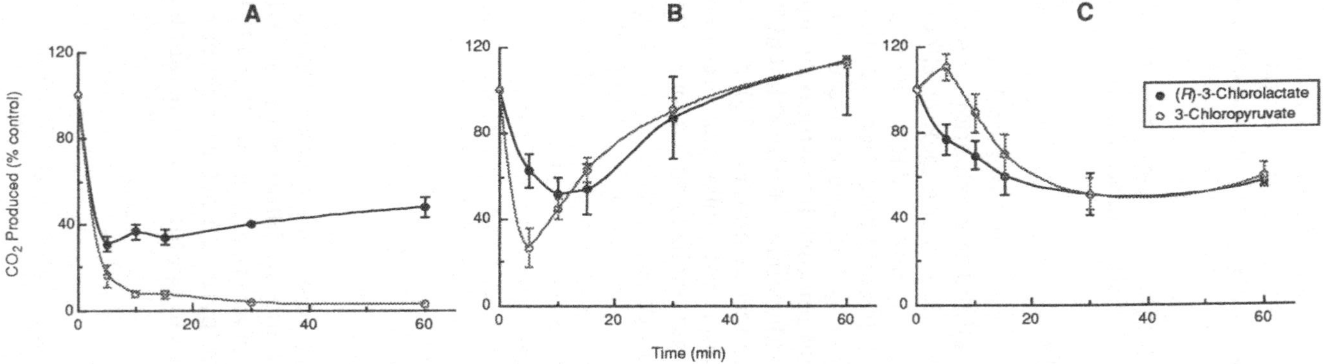

Figure 20.1 The effect of (*R*)-3-chlorolactate (5 mM) and 3-chloropyruvate (1 mM) on the production of $^{14}CO_2$ from: (A) D-[U-^{14}C]fructose (1 mM); (B) [U-^{14}C]glycerol (2 mM) and (C) L-[U-^{14}C]lactate (2mM). Washed spermatozoa suspended in phosphate buffered saline were incubated in the presence of each combination of substrate and the sodium salt of the inhibitor at 34°C in sealed Warburg flasks for the times indicated. Each point represents the mean ± S.E.M. (n = 3 to 6).

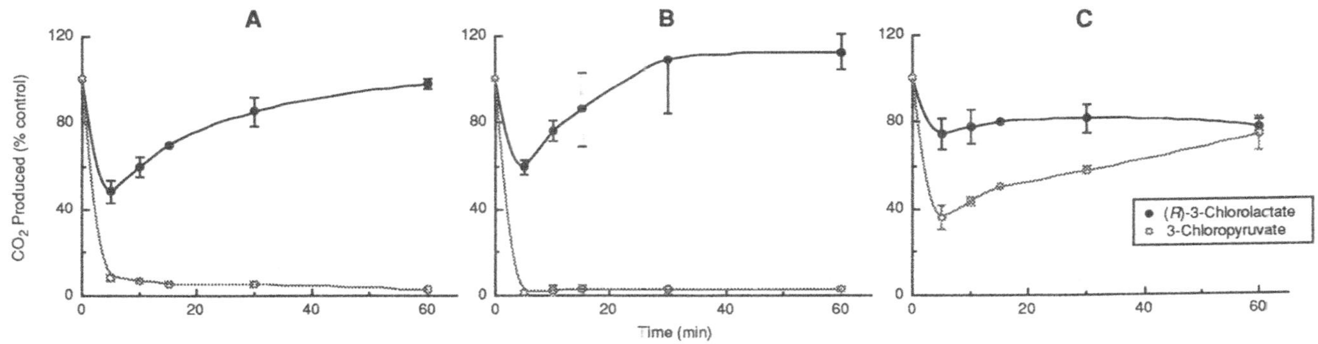

Figure 20.2 The effect of (*R*)-3-chlorolactate (5 mM) and 3-chloropyruvate (1 mM) on the production of $^{14}CO_2$ following a 10 min incubation of cells in the presence of either inhibitor before the addition of radiolabeled substrate. Substrates and experimental conditions are as indicated in Figure 20.1.

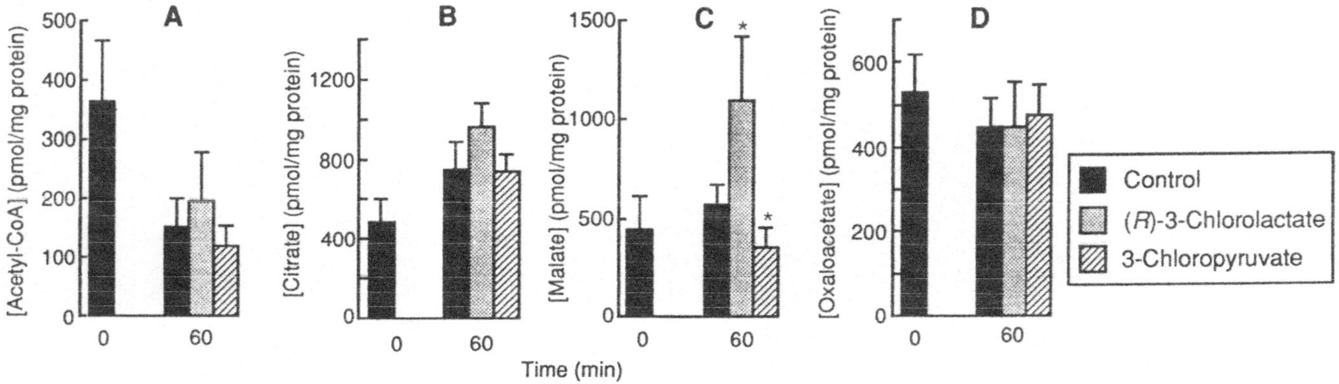

Figure 20.3 The effect of (*R*)-3-chlorolactate (5 mM) and 3-chloropyruvate (1 mM) on the concentrations of (A) acetyl-CoA, (B) citrate, (C) L-malate and (D) oxalacetate in boar spermatozoa following 1 h incubation with L-lactate (2 mM). Mean ± S.E.M. (n = 6 to 8). * = $p < 0.10$.

REFERENCES

Cooney, S.J. and Jones, A.R. (1988) Inhibitory effects of (S)-3-chloroacetaldehyde on the metabolic activity of boar spermatozoa *in vitro*. *J. Reprod. Fertil.* **82**, 309–317.

Dobbie, M.S., Porter, K.E. and Jones, A.R. (1988) Is the nephrotoxicity of (R)-3-chlorolactate in the rat caused by 3-chloropyruvate? *Xenobiotica* **18**, 1389–1399.

Jones, A.R. and Chantrill, L. (1989) Oxidative metabolic activity of boar spermatozoa: a system for assessing anti-glycolytic activity of potential inhibitors *in vitro*. *Reproduction, Fertility and Development (in press)*.

Jones, A.R., Mashford, P.M. and Murcott, C. (1979) The metabolism of 3-amino-1-chloropropane-2-ol in relation to its antifertility activity in male rats. *Xenobiotica* **9**, 253–261.

Porter, K.E. and Jones, A.R. (1982) The effect of the isomers of alpha-chlorohydrin and racemic beta-chlorolactate on the rat kidney. *Chem.-Biol. Interactions* **41**, 95–104.

21 BIOMONITORING OF DNA-DAMAGING TOXINS

GARY T. VAUGHAN AND T. MARK FLORENCE

INTRODUCTION

Many common types of cancer are linked to exposure to environmental agents, both man-made and naturally-occurring. The human diet contains substantial amounts of a wide variety of mutagens and carcinogens. The hydrazine derivatives in mushroom species, pyrrolizidine alkaloids in many species of plants, theobromine in tea and cocoa, chlorogenic acid in coffee, and mycotoxins such as aflatoxin are known mutagens and animal carcinogens. Other carcinogens are produced by the pyrolysis of foods and other organic material. In addition to these naturally-occurring carcinogens, man has introduced many industrial chemicals, pesticides and drugs which have been linked to human cancer.

These substances come from a variety of chemical groups including polyaromatic hydrocarbons, aromatic amines, nitrosamines, halohydrocarbons and alkylating agents. Although these chemicals are not structurally related they all are, or are metabolized to, electrophilic forms which react with nucleophilic sites in DNA, RNA and proteins to form stable addition products called adducts. If DNA adducts are not repaired, hereditable changes in genetic material can occur which can lead to the induction of tumors or hereditable genetic disease.

There is increasing concern about exposure to these genotoxins, particularly in the workplace. Because there is often a lag period of many years for diseases such as cancer to develop, individuals may be unaware that they were exposed to excessive levels of these substances until many years after exposure began.

Therefore, there is a need for the detection of excessive exposure at the time of occurrence when a significant reduction in the risk to health can be achieved by a reduction or termination of exposure.

In the workplace, exposure to genotoxic chemicals has traditionally been controlled by setting standards for the concentration of compounds in ambient air and monitoring the air to ensure that these levels are not exceeded. However, this only takes into account exposure via the pulmonary route and thus underestimates total exposure especially for substances that are mainly absorbed through the skin. By contrast, biological monitoring of exposure by measuring the concentrations of genotoxins or their breakdown products in body fluids or excreta takes into account absorption by all routes and integrates exposure from all sources. The determination of genotoxins in body fluids gives information of the amount of the compound that is absorbed but it does not estimate the extent of the formation of DNA adducts, a critical event in tumor initiation. However, the direct determination of DNA adducts is complicated by the small amounts available in accessible tissue such as white blood cells and by the fact that DNA adducts are removed by chemical and enzymatic processes at different rates.

Unlike DNA adduct, protein adducts are not subject to repair and are often stable for the life of the protein. Haemoglobin has been shown to be particularly useful for monitoring the internal dose of genotoxic chemicals (Neumann, 1984). It is readily available in gram amounts from a sample of blood and its relatively long life-span (120 days) permits the determination of cumulative doses of genotoxic chemicals over a period of months. Studies with a variety of genotoxic chemicals have shown a proportionality between the formation of specific products of reaction with DNA and haemoglobin (Ehrenberg and Osterman-Golkar, 1980). The rates of formation of haemoglobin adducts, therefore, provide an indirect measurement of the rates of formation of the corresponding DNA adducts.

The concentrations of haemoglobin adducts are usually measured as fmol to pmol per gram of haemoglobin and the measurement of these low concentrations requires methods that are extremely sensitive and selective. The aim of the present work was to determine the exact positions of adducts in the haemoglobin molecule and to use this structural information to optimize analytical procedures for the determination of the adducts.

METHODS AND RESULTS

Whole blood (1 ml) was methylated *in vitro* with 5.5 MBq [^{14}C]methyl methanesulphonate [2.15 GBq/mmol] at 37°C for 2 h. The proportion of radioactivity bound to the α- and β-chains of haemoglobin was determined by chromatographing 1/1000 of the red cell haemolysate on a C_4 HPLC column and determining the radioactivity in 1 ml fractions by scintillation counting (Figure 21.1) The α- and β-chains contained 42 and 58% of the radioactivity bound to haemoglobin respectively.

[^{14}C]Methyl haemoglobin was isolated by precipitation from the red cell haemolysate with 10 volumes of acetone containing 1% HCl. The haemoglobin contained 10.2% of the radioactivity which was added to the blood. Ten mg haemoglobin was digested with trypsin (100 mg) in 2% ammonium bicarbonate at pH 8 for 24 h at 37°C. The tryptic peptides were separated by reversed phase HPLC and radioactivity was detected with a continuous-flow radioactivity monitor (Figure 21.2). The major products were identified as the N-terminal peptides (α-1 and β-1) by their retention time and by amino acid analysis on HPLC after precolumn derivatization with o-phthaladehyde. Minor products were due to the methylation of histidine residues in various peptides. Approximately half of the radioactivity was found in the insoluble

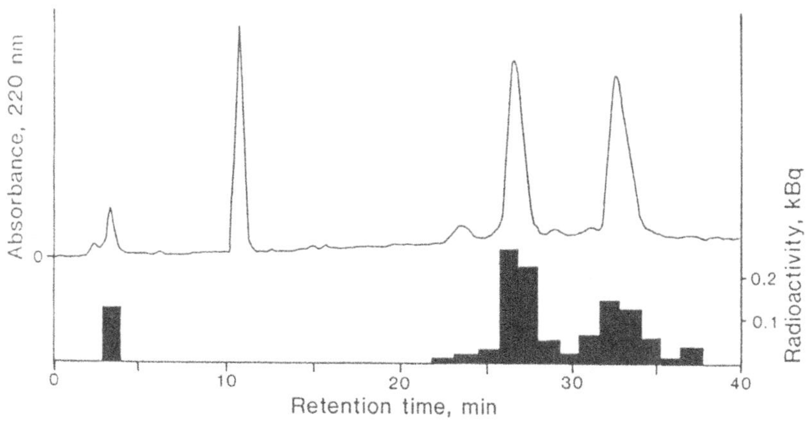

Figure 21.1 HPLC separation of α- and β-chains of [^{14}C]methyl hemoglobin. Column: 4.6 x 250 mm Vydac C_4. Mobile phase: linear gradient from 40–50% acetonitrile in water (0.1% TFA) at 1 ml/min. Heme (11 min), β-globin (27 min), α-globin (33.5 min).

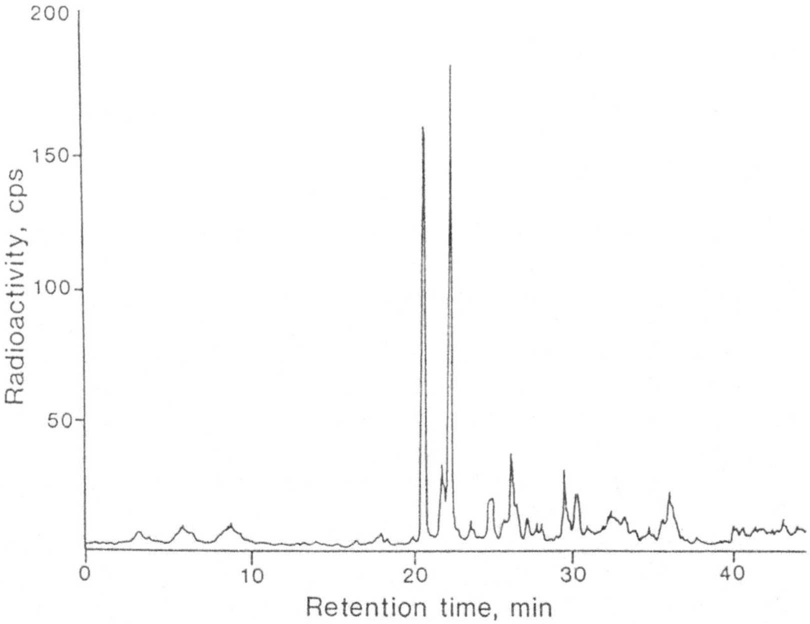

Figure 21.2 Separation of tryptic peptides of [^{14}C]methyl hemoglobin by HPLC. Column: 4.6 x 250 mm Vydac C_{18}. Solvent A: water (0.1% TFA). Solvent B: acetonitrile (0.1% TFA). Linear gradients: 0–35% B in 30 min, 35–60% B in 5 min, 60–0% B in 2 min at 1 ml/min. α-1 (20.5 min), β-1 (22.5 min).

core peptides and is probably due to adducts with cysteine residues˙

The N-terminal adducts can be determined using a modified Edman degradation procedure (Törnqvist *et al.*, 1986), in which alkylated N-terminal adducts are selectively removed in a neutral-alkaline coupling medium. Unsubstituted N-terminal amino acids, which require acidic conditions for cleavage to occur, remain attached to haemoglobin. When pentafluorophenyl isothiocyanate is used as the coupling reagent, background levels of N-methyl valine and N-2-hydroxyethyl valine (from exposure to ethylene oxide) were detected by gas chromatography-mass spectrometry (GC-MS). The F_5PTH amino acids were separated on a BP-5 capillary column and detected by methane chemical ionization mass spectrometry and negative-ion detection. Methyl valine- and 2-hydroxethyl valine-F_5PTHs were also determined by GC with electron capture detection and HPLC on a 5 m C_{18}

column with 45% acetonitrile in 30 mM sodium acetate, pH 5 at 2 ml/min and UV detection at 269 nm. However, both methods were prone to interference at low levels of adducts although this was improved by pre-purification on C_{18} and silica Sep-Pak cartridges.

CONCLUSIONS

Methylation of the primary amino group of the N-terminal amino acid of haemoglobin was a major product of reaction with methyl methanesulphonate. Unlike other genotoxicants, such as 4-aminobiphenyl which preferentially alkylates the β-93 cysteine residue of haemoglobin, methyl methanesulphonate reacts with both α- and β-globins almost equally. The modified Edman degradation procedure of Törnqvist *et al.* (1986) provides a useful means of isolating and measuring the concentrations of these adducts.

REFERENCES

Ehrenberg, L. and Osterman-Golkar, S. (1980) Alkylation of macromolecules for detecting mutagenic agents. *Teratogen. Carcinogen. Mutagen.* **1**, 105–127.

Neumann, H.-G. (1984) Analysis of haemoglobin as a dose monitor for alkylating and arylating agents. *Arch. Toxicol.* **56**, 1 6.

Törnqvist, M., Mowrer, J., Jensen, S. and Ehrenberg, L. (1986) Monitoring of environmental cancer initiators through haemoglobin adducts by a modified Edman degradation method. *Anal. Biochem.* **154**, 255–266.

22 ORGANOCHLORINES—POLLUTANTS AND HUMAN CARCINOGENS

MICHAEL McCABE, DAVID MAGUIRE AND DEBRA YIN-FOO

INTRODUCTION

In 1975 Fridovich proposed that superoxides are the basis for oxygen toxicity. Since then the generation and the fate of oxygen derived free radicals has been increasingly detailed, because of the suspicion that these compounds may play a role both in the genesis of some human cancers, as well as be intimately involved in the ageing process. It is known that the unregulated production of hydroxyl radicals can cause massive cellular damage, and it is perhaps for this reason that mammals severely restrict the presence of ferrous and cuprous ions within the body by complex formation within a series of carriers (transferrin/ferritin and lactoferrin for Fe^{2+} and caeruloplasmin/albumin for Cu^{2+}), since these metal ions would otherwise participate in Haber-Weiss type Fenton reactions to generate OH from any metabolically formed hydrogen peroxide.

During the mid 1970s a suspicion emerged that the incidence of human adult leukemias might be elevated in sugar cane growing regions in Queensland. An epidemiological study (McCabe *et al.*, 1984) has since revealed that this is so. The sugar cane industry in Australia is a prolific consumer of a variety of second generation pesticides including the organochlorines. However the carcinogenic role of organochlorines in humans is equivocal since they are not generally mutagenic when subjected to conventional (Ames type) mutagenicity tests. Proof that a chemical (even one known to be mutagenic and to cause tumors in experimental animals) is also carcinogenic in man, is notoriously difficult to obtain as it would require extremely complex and lengthy epidemiological studies. The epidemiological data so far gathered in the Burdekin (sugar cane growing region of northern Queensland) shows that the leukemia mortality is significantly increased only in cane farmers over 60 years (see Table 22.1), a finding which is certainly consistent with an occupational exposure to a chemical carcinogen with perhaps a long latency. In these circumstances the demonstration of a plausible carcinogenic mechanism for the suspected chemical(s) is important even though it cannot be a direct proof of the involvement of the compound in the genesis of the cancer.

The class of organochlorine pesticides are characterized by their known effects upon the functioning of the CNS in insects, which is thought to be due to their severe inhibition at the micromolar level of a Ca^{2+} and Mg^{2+} dependent ATP'ase (Lee and Park, 1979). Since it is known that Ca^{2+} is a most important second messenger in the commitment to cell proliferation of the cells of the immune system (Freedman, 1979), a role for organochlorines in the modulation of (immune cell) calcium signals has been investigated. Additionally the possibility that the cells of the immune system may "detoxify" organochlorines, perhaps with the

Table 22.1 Number of deaths and standardized mortality ratios for leukemias for the period 1960–1981

		1960–1967		*1968–1981*	
		Men	*Women*	*Men*	*Women*
Townsville					
	All ages	16	8	21	23
		(101)	(68)	(62)*	(98)
	>60 years	6	5	10	10
		(81)	(82)	(57)	(85)
Cane-growing areas					
	All ages	28	27	86	31
		(79)	(102)	(119)	(62)*
	>60 years	13	15	61	13
		(80)	(112)	(154)**	(49)

Standardized mortality ratios in parentheses.
* $P < 0.05$; ** <0.01

concomitant generation of unscheduled free radical oxygen has also been investigated.

METHODS

Preparation of bovine lymphocytes was from heparinized blood samples using standard methods involving a Ficoll-Pacque gradient (McCabe *et al.*, 1983).

Thymus and peritoneal cells were obtained from 7 week female outbred Wistar rats by standard techniques from minced and sieved thymus glands, while peritoneal cells were washed from the peritoneal cavity using phosphate buffered saline containing heparin. All cell suspensions were washed by centrifugation and the viability of cells assessed by staining with Trypan Blue.

Calcium uptake was measured from triplicate samples of cells previously incubated for an hour in the presence of the organochlorine using ^{45}Ca. The cells were centrifuged through a silicon oil layer to separate them from the incubation medium, and cell numbers were assessed by counting $^{3}H_2O$ added to the medium and which accompanied the lymphocytes through the oil phase (McCabe and Yin-Foo, 1986).

Chemiluminescence was measured using a scintillation counter set into "out of coincidence" mode. Light emission was amplified by the inclusion of luminol which reacts with reactive oxygen species (produced during the stimulation of phagocytic cells) to produce an excited aminophthalate ion which releases photons during its return to a ground state.

RESULTS AND DISCUSSION

The addition of an organochlorine (DDT) to a suspension of lymphocytes caused a significant dislocation of Ca^{2+} uptake even at 10 mmolar (see Figure 22.1). Similar effects were also shown by micromolar concentrations of a variety of other organochlorines including dieldrin and chlordane.

The addition of organochlorines to thymus cells also perturbed chemiluminescent responses of the cells to a variety of stimuli and most particularly when the calcium ionophore A23187 was also added. Figure 22.2 shows the chemiluminescent response of rat thymus cells stimulated with both A23187 and chlordane. The simultaneous addition of these two compounds resulted in a rapid and bimodal stimulation of chemiluminescence. A similar

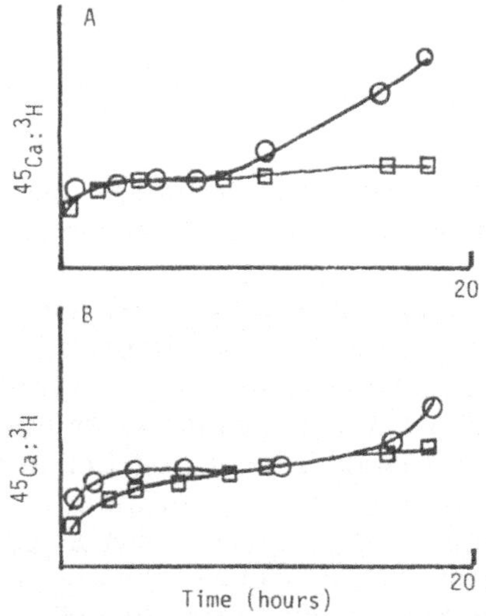

Figure 22.1 Effect of DDT on Ca^{2+} uptake in Con A-stimulated bovine lymphocytes, in the presence (o) and absence (□) of Con A (100 μg/ml). A without DDT, and B in the presence of DDT at 10^{-5}M.

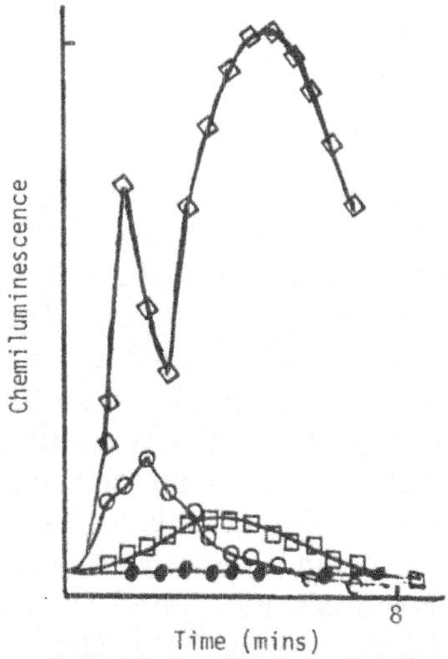

Figure 22.2 The chemiluminescent response of rat thymus cells stimulated with A23187 and chlordane in the presence of luminol. (◊) 10 μM A23187 and 100 μM chlordane; (o) A23187 alone; (□) chlordane alone.

large and synergistic chemiluminescent response was obtained using rat peritoneal cells, although in this case the bimodal expression of chemiluminescence was not apparent (Yin-Foo, 1987).

A combination of centrifugal counterflow elutriation and isopycnic separation techniques using discontinuous Percoll gradients has shown that the chemiluminescent activity copurifies consistently with the neutrophil enriched fractions of both the thymus and peritoneal cavity cell mixtures, and is almost completely absent from the macrophage/monocyte and lymphocyte bands. Identification of the cell population with enriched chemiluminescent activity as being a neutrophil fraction was based upon its functional as well as morphological characteristics, for example staining characteristics and cell size, morphology and density, as well as the tendency to adhere to glass or plastic surfaces.

By the addition of a variety of scavengers and inhibitors, attempts were made to identify the nature of the oxygen species generating the chemiluminescent response. The addition of superoxide dismutase at 120 μg/ml when incubated with thymus cells, strongly inhibited both peaks of the bimodal response, each by approximately 50%. The addition of ascorbic acid also caused a similar degree of inhibition. These results suggest the involvement of O_2 or of O_2 derived compounds in the generation of the chemiluminescence. The incompleteness of the inhibition suggests that the product of dismutation (i.e. H_2O_2) may be responsible for generating a part of the signal (Yin-Foo, 1987).

The metabolic transformations of xenobiotics generally are performed by enzymes of the endoplasmic reticulum. For hydrophobic compounds these metabolic transformations often depend upon the activities of a cytochrome P_{450} dependent system of monooxygenases which are present in significant amounts in the liver microsomal and the adrenal mitochondrial fractions. It is known that even brief exposure to organochlorine pesticides will induce a significant increase in the cytochrome P_{450} content of liver and adrenals in mammals. The other major microsomal monooxygenase system is the FAD containing monooxygenase (Kulkarni and Hodgson, 1984) which is known to be the mechanism for sulphoxidation, desulphuration and amine oxidation. Presumably both oxygenase systems may be involved in any particular xenobiotic detoxification. In addition to these systems, Eiling *et al.* (1983) have shown that the co-oxidation of xenobiotics can occur during the synthesis of prostaglandins. The prostaglandin synthase is located in the endoplasmic reticulum and includes a hydroperoxidase activity. It is thought to be this activity which is responsible for the co-oxidation of xenobiotics (Marnet, 1981; Moldeus *et al.*, 1985).

There are thus several possible metabolic routes whereby organochlorines may generate superoxide in neutrophils, and in order to decide the metabolic source of the superoxide, whole cell sonicates were prepared and the various cell constituents were separated by density gradient centrifugation using a Percoll gradient. Particulate fractions were identified by electron microscopy and by assay of marker enzymes such as myeloperoxidase. Each of the isolated fractions was investigated for its ability to generate chemiluminescence in response to chlordane and A23187. The results suggest that the intracellular location of the radical generating reactions is in the microsome rich fraction of the neutrophil, and takes place largely through the cytochrome P450 system, although this does not preclude the possibility of a contribution through the co-oxidation of organochlorines with prostaglandin synthesis.

Lipoproteins are the principle carriers of organochlorine pesticides in the body and will deposit them into cells through the LDL uptake system of receptor mediated pinocytosis, as is seen in lymphocytes (Goldstein and Brown, 1977). While the respiratory burst of neutrophils and other phagocytic cells does generate a large liberation of reactive oxidants, these are specifically released in the close proximity of, and in a directed manner towards tissue designated for destruction (invading microorganisms, tumor cells etc.). Thus phagocytic cells do not normally generate oxidizing free radicals in a random or unscheduled manner.

The production of such unscheduled oxidizing free radicals is known to be extremely damaging to surrounding cells and tissues and if generated by organochlorines, would be occurring in tissues already stressed by calcium imbalance. Such an upset in calcium signals would normally push cells back into a G_0 (resting) state. However we have shown (McCabe and Yin-Foo, 1986) that some fraction at least of a lymphocyte population can nevertheless be forced into S-phase and initiate cell divisions. Such cells have apparently accommodated to the absence of normal regulatory controls by calcium and in this sense they have begun to resemble cancer cells (Hickie *et al.*, 1982).

REFERENCES

Eiling, T.E., Reed, G., Boyd, J., Krauss, R.S. and Sivarajah, K. (1983) Metabolism of chemical carcinogens by prostaglandin H synthase. *Drug Metab. Rev.* **14**, 219–248.

Freedman, M.H. (1979) Early biochemical events in lymphocyte activation. Investigations on the nature and significance of early calcium fluxes observed in mitogen induced T and B lymphocytes. *Cell Immunol.* **44**, 290–313.

Fridovich, I. (1975) Superoxide dismutases. *Ann. Rev. Biochem.* **44**, 147–149.

Goldstein, J.L. and Brown, M.S. (1977) The low density lipoprotein pathway and its relation to atherosclerosis. *Ann. Rev. Biochem.* **46**, 897–930.

Hickie, R.A., Wei, J., Blyth, L.M. and Wong, D.W. (1983) Cations and calmodulin in normal and neoplastic cell growth regulation. *Can. J. Biochem. Cell Biol.* **61**, 934–941.

Kulkarni, A.P. and Hodgson, E. (1984) Metabolism of insecticides by mixed function oxidase systems. *Ann. Rev. Pharmacol. Toxicol.* **24**, 19–42.
Lee, T.P. and Park, B.H. (1979) Biochemical basis of aroclor 1254 and pesticide toxicity *in vitro*. *Res. Comm. Chem. Pathol. Pharmacol.* **25**, 597–605.
McCabe, M., Maguire, D. and Nowak, M. (1983) The effects of arsenic compounds on human and bovine lymphocyte mitogenesis. *Environ. Res.* **31**, 323–331.
McCabe, M., Nowak, M., Hamilton, R. and Hartshorn, P. (1984) Cancer of lymphatic tissues in cane-growing areas of Queensland. *Med. J. Aust.* **141**, 412–414.
McCabe, M. and Yin-Foo, D. (1986) Effects of DDT on the calcium transport and thymidine uptake of bovine lymphocytes. *Bull. Environ. Contam. Toxicol.* **37**, 523–530.
Marnet, L.J. (1981) Polycyclic aromatic hydrocarbon oxidation during prostaglandin biosynthesis. *Life Sci.* **29**, 531–546.
Moldeus, P., Ross, D. and Larsson, R. (1985) Co-oxidation of xenobiotics. *Biol. Soc. Trans.* **13**, 847–850.
Yin-Foo, D. (1987) Oxidant generation in immune cells upon interaction with organochlorine pesticides. PhD thesis, James Cook University.

23 ECOTOXICOLOGY

DES W. CONNELL

INTRODUCTION

Environmental issues are now matters for widespread public concern and among these the management of chemicals is a high priority issue. The control of chemical discharges ranks high in the business of the U.S. Congress (Anon., 1989) and other governments throughout the world. Environmental Chemistry (Connell, 1989) is now a major area of professional activity in government and industry. At the international level the OECD has taken a major initiative in co-ordinating the control of chemicals (Connell, 1986).

About 100 000 chemicals are estimated to be in daily use as antibiotics and other pharmaceuticals, plastics, food preservatives, pesticides and so on (Freed, 1980). Most of these chemicals are beneficial to society and have been used without any adverse effects on the natural environment or human health. However there are some which have had adverse effects. The need to develop a capability to evaluate the environmental effects of chemicals has arisen principally from these situations.

SCOPE OF ECOTOXICOLOGY

Toxicology is concerned with the effects of chemicals on biota. A great deal of work has been carried out on the toxic effects on particular species of mammals, fish and other organisms of specific interest. A large body of information is now available on lethal and other effects of chemicals on these organisms. However the extrapolation of this information to the effects of a chemical on a complex natural ecosystem is limited. The exposure of biota in the system to the chemical is not known.

Also the effects of the complex array of interacting organisms present in a natural ecosystem cannot be determined. This has led to the development of *ecotoxicology* which provides a framework within which these problems can be evaluated (Butler, 1978; Connell and Miller, 1984).

The broad scope and approach of ecotoxicology can be diagrammatically illustrated as in Figure 23.1. The source of the chemical is identified and quantified, as are the distribution, transport and transformation processes whereby the response of individuals in the environment can be evaluated. Organism response can be assessed to some extent, using toxicological data and finally the effects of population, community and ecosystem considered.

SOURCE OF CHEMICALS

Chemicals can be discharged to the environment in a wide variety of ways, as point discharges from an industrial operation which uses or manufactures chemicals or in a dispersed form from useage or disposal.

To evaluate the extent of possible distribution of a chemical in the environment, data are needed on possible production quantities and location, intended use, transport methods, and final disposal of any wastes involved. From this, particular sections of the environment that may be affected can be identified and the quantities and concentrations of the chemical likely to be discharged calculated.

ENVIRONMENTAL DISTRIBUTION PROCESSES

The distribution of chemicals in the environment is a highly complex process but is governed by the physicochemical properties of the chemicals and the environment itself. As a first step in understanding these processes the environment can be simplified by considering it to be derived into separate phases or compartments within which a chemical behaves in a uniform manner. The phases most commonly used are air, soil, water, sediments and biota (see Figure 23.2).

It is possible to subdivide some of these phases further, for example sediments into bottom and suspended sediments, if further refinement is required, and if data are available on the phases and the behavior of the chemical in it. Clearly this ap-

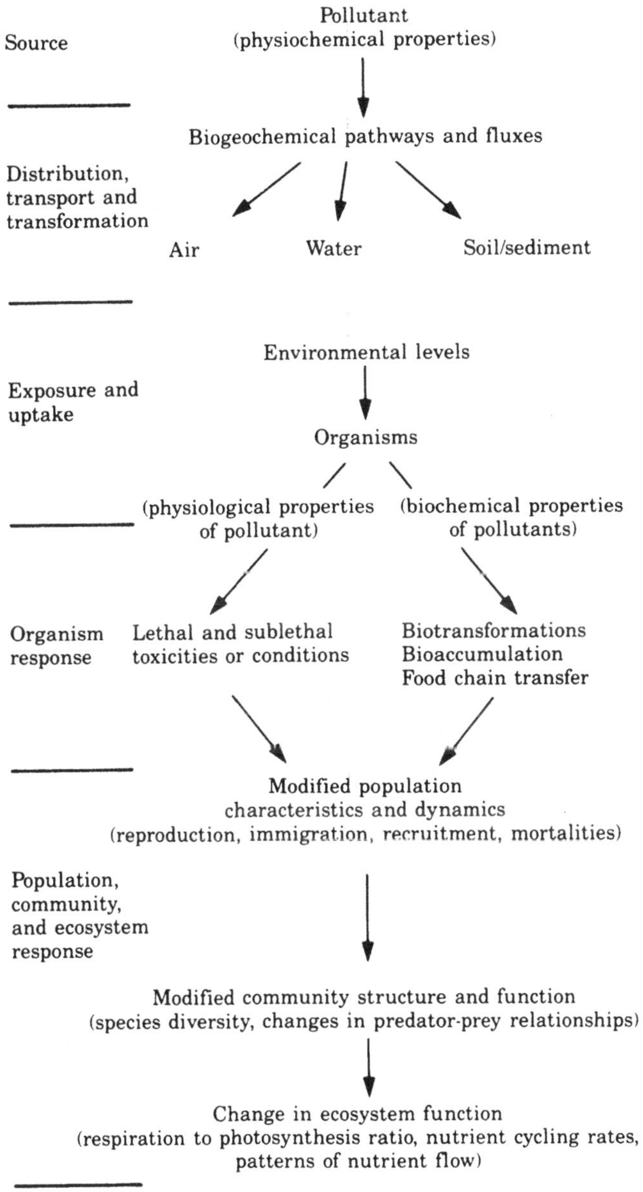

Figure 23.1 Diagrammatic illustration of the impact of pollutants on the components and functions of natural ecosystems with some important characteristics causing or modified by, change indicated in parenthesis (from Connell and Miller, 1984).

proach involves some over-simplification but if more basic data are available the phases used can be made to reflect the actual environment more accurately.

When a chemical is discharged to the environment it partitions between the various phases present as shown in Figure 23.2. This partitioning is characterized by the distribution or partition coefficient, K_{aso}, K_{sew}, etc.

Next the properties of the chemical which influence its environmental distribution need to be considered. The property of most concern is the 'fugacity' (Mackay and Patterson, 1981). This is often described as the 'escaping tendency' of a substance from a phase. It can be accurately calculated but has different characteristics in different phases with the same compound. Fugacity in a particular phase is usually linearly related to concentration at low concentrations such as are encountered with environmental contaminants. Thus

$$f = C/Z$$

where f is the fugacity, C the concentration, and Z the fugacity capacity or constant. So to calculate the fugacity of a compound the fugacity constants must be known and these can be derived from data on physicochemical properties such as the Henry's Law constant, the soil-to-water distribution coefficient and so on.

The great value of fugacity is that at equilibrium the fugacities in all the different phases are equal, so

$$f_{equil} = f_{air} = f_{water} = f_{sediments} = f_{biota}$$

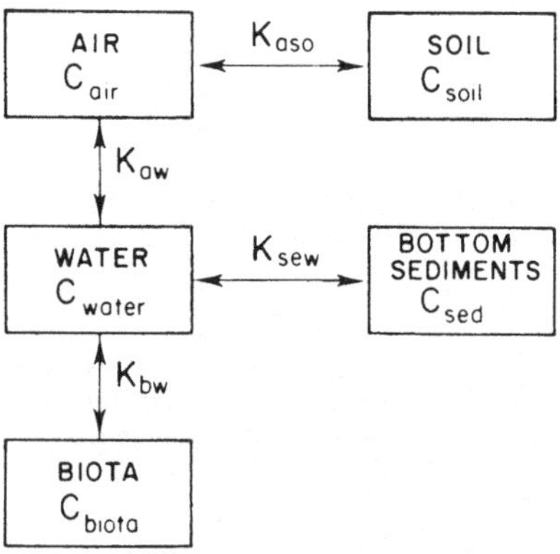

Figure 23.2 Environmental phases and their interrelationships for a chemical discharge to the environment.

Thus there is one value for fugacity (f_{equil}) which applies to the compound as it distributed in each phase. It can be shown that

$$f_{equil} = C_i/Z_i = M_i/V_iZ_i = \Sigma M_i/\Sigma (V_iA_i) = M_{total}/\Sigma (V_iZ_i)$$

where M_i is the number of moles of chemical in a phase of volume V_i, C_i the concentration of the chemical and Z_i, the fugacity capacity constant.

Using this equation the prevailing equilibrium fugacity can be determined and, from this, the amount and concentrations of the substance in each individual phase can be calculated.

However, to calculate f, the total mass involved (M_{total}) expressed as moles must be known, the volume of each phase (V_i) measured and the fugacity capacity constants (Z_i) of the substance in each phase estimated. The fugacity capacity constants can be estimated from basic physicochemical data (Mackay and Patterson, 1982).

Connell and Hawker (1986) have used this modelling method and additional phases, from those illustrated in Figure 23.1, of terrestrial vegetation and suspended sediments. Their phase characteristics, developed for Canberra, are shown in Table 23.1.

The potential distributions and concentrations of two chemicals are shown in Table 23.2. In the Australian environment different volumes would be required to fit different regions. Of course the model assumes that the phases are homogeneous, which is often a reasonable assumption with air but can be less

Table 23.1 Phase characteristics for Canberra, Australia

	Volumes for Canberra are (m^3)	*Expressions for Z ($mol\ m^{-3}\ Pa^{-1}$)*
Atmosphere	2.21×10^{11}	1/RT
Water	2.32×10^{7}	1/H
Soil	4.32×10^{6}	$(0.63\ K_{ow} \times OC \times \rho_s)/H$
Suspended Sediment	2.99×10^{3}	$(0.63\ K_{ow} \times OC \times \rho_{ssed})/H$
Sediment	2.11×10^{5}	$(0.63\ K_{ow} \times OC \times \rho_{sed})/H$
Aquatic biota	1.66×10^{1}	$K_{ow} \times L_b\ \rho_b)/H$
Vegetation	9.67×10^{4}	$K_{ow} \times L_v\ \rho_v)/H$

From: Connell and Hawker, 1986

K_{ow} = octanol/water partition coefficient   L = lipid fraction

H = Henry's law constant   OC = organic carbon

Table 23.2 Calculated potential distribution and concentration of representative chemicals in the Canberra environment

	Tetrachloroethene		*DDT*	
	Con μg/kg	*% moles*	*Con μg/kg*	*% moles*
Air	0.5	82.7	5×10^{-6}	3×10^{-4}
Water	2×10^{-3}	0.03	5×10^{-6}	4×10^{-5}
Soil	3.8	14.9	47	86.5
Suspended sediments	11.4	0.03	141	0.18
Sediment	11.4	2.17	141	12.6
Aquatic biota	30	3×10^{-4}	376	2×10^{-3}
Vegetation	11.9	0.10	150	0.61

applicable with water and soil. The greater the variability, the less accurately the model will reflect the actual environment.

This model provides a method for predicting environmental concentrations for a single discharge of non-degradable chemical into a static environment. This type of model provides a valuable basis for comparison of chemicals and identifying the sectors of the environment which are likely to be repositories of discharged chemicals.

However if estimates of actual concentrations likely to occur are required then additional factors need to be taken into account. For example allowance is needed for the rate of discharge of a chemical and its rate of degradation in different phases.

Also the rate of movement of water and air phases results in the removal of chemicals from the model boundaries. All of these factors can be incorporated into the model if the relevant data is available. The accuracy of the results obtained is related to how well the data reflect the actual environment.

BIO-ACCUMULATION BY ORGANISMS

The bio-accumulation of chemicals in the environment by organisms is of particular importance. This process controls the concentrations of contaminant chemicals in organisms and thus the deleterious effects on biota. In general, organisms take up chemicals from two principal sources, directly from the ambient environment and from food. The atmosphere contains very low concentrations of contaminants compared with water bodies, such as the oceans, lakes, rivers etc., and as a result the water to organism relationships have received most attention.

The bio-accumulation process can be seen as a balance between two kinetic processes, uptake and clearance (Moriarty, 1985).

The concentration in an organism will approach equilibrium when the rates of uptake and clearance are equal. The bio-accumulation factor (K_{bw}) is the ratio of concentration in the biota to concentration in the environment. For a particular compound and particular organisms, or groups of organisms, e.g. fish, K_{bw} has been shown to approach a constant value (Mackay, 1982). A knowledge of the bio-accumulation factor and the concentration of a chemical in the environment will allow estimation of the concentrations likely to occur in organisms.

Connell (1988) has pointed out that for lipophilic (fat soluble) compounds the major concentrating phase in biota is the lipid phase and the aqueous tissues have little effect. Thus bio-accumulation can be viewed as essentially the establishment of an equilibrium between biota lipid and surrounding factor can be related to the octanol (lipid-like) to water partition coefficient, K_{ow}.

The relationship of these factors takes the following general form

$$\log K_{bw} = n \log K_{ow} + b$$

where n and b are empirical constants.

The K_{ow} value can be accurately and simply measured in the laboratory and thus provides a means of evaluating bioconcentration.

RESPONSE OF ORGANISMS

The assimilation of chemicals by organisms may induce a range of biological effects from the sublethal to the lethal level. Investigations of the lethal effects of environmental contaminants is the basic approach presently used in environmental toxicology. Laboratory exposure of a uniform set of organisms, under controlled laboratory conditions, to a toxicant at different concentrations allows the relationship between percentage lethality and toxicant concentration to be established. This LC_{50} data can be compared directly with concentration calculated to occur in a given situation using the procedures previously outlined.

However there are some major problems with this approach. For example the laboratory conditions and the test animals may not be representative of those in the natural environment. Also only a limited number of species can be evaluated by this procedure compared with the wide range of species present in natural ecosystems. Nevertheless, the possibility of lethal effects with some organisms at concentrations comparable to, or above, the established lethal levels can be established. Sublethal effects also present a variety of problems in evaluating their environmental significance.

ECOSYSTEM RESPONSE

Organisms in an ecosystem exhibit differing susceptibilities to toxicants. For example, herbicides are selectively toxic to plants, and insecticides have a major impact on insect populations. Even within these broad groups of organisms there are a wide range of different susceptibilities with different species. Thus when any toxicant enters an ecosystem it will selectively remove those susceptible species from the range of organisms present. This generally leads to a reduction in the species diversity of organisms present and a change in the community structure.

In addition to toxic effects on different components of an ecosystem, other important secondary effects usually result. In natural ecosystems sets of animals and plants have adapted to the existing environmental conditions. There a wide range of interacting and interdependent organisms including microorganisms, plants, invertebrates, vertebrates and so on. But the interrelationships between these groups are usually not well understood, which can lead to difficulties in understanding the effects of toxicants. However, in many cases the effects due to a toxicant are large compared with natural variations and can lead to a more obvious ecological effect.

CONCLUSIONS

Ecotoxicology has developed considerably in recent years stimulated by governmental requirements for the registration of new chemicals. Methods for evaluation of the behavior of chemicals in the environment have been developed which are relatively simple and provide reasonably accurate results.

However, there is a great deal further to be done, particularly with regard to the effects on ecosystems. In evaluating the current information available it seems there is no existing or new persistent organic chemical for which the full range of significant effects on ecosystems can be predicted at sublethal concentrations over extended periods of time in the natural environment.

Even with the completion of all testing required by present assessment programs there is still little knowledge of the effects at different life stages, sublethal effects, synergistic effects with other pollutants and so on. Thus monitoring of behavior and effects of chemicals in use is also needed.

REFERENCES

Anon. (1989) Congress tussles with environmental issues. *Chem. Eng. News*. **67**, 18.

Butler, G.C. (1978) Principles of Ecotoxicology, Scope 12, New York: John Wiley and Sons.

Connell, D.W. (1986) Ecotoxicology—a new approach to understanding hazardous chemicals in the environment. *Search* **17**, 27.

Connell, D.W. (1988) Bioaccumulation behaviour of persistent organic chemicals with aquatic organisms. *Rev. Environ. Contam. Toxicol.* **101**, 117.

Connell, D.W. (1989) Environmental Chemistry—a new branch of chemistry. *Chem. Aust.* **56**, 264.

Connell, D.W. and Miller, G.J. (1984) *The Chemistry and Ecotoxicology of Pollution*, New York: John Wiley and Sons.

Freed, V.H. (1980) The University Chemist—Role in Transport and Fate of Chemical Problems. In: *Dynamics, Exposure and Hazard Assessment of Toxic Chemicals*, edited by R. Hague. p.27. Ann Arbor: Ann Arbor Science.

Mackay, D. and Patterson, S. (1982) Fugacity revisited. *Environ. Sci. Technol.* **16**, 654A–660A.

Moriarty, F. (Ed) (1975) *Organochlorine Insecticides Persistent Organic Pollutants*. London: Academic Press.

INDEX

For Product Safety Concerns and Information please contact our EU representative GPSR@taylorandfrancis.com Taylor & Francis Verlag GmbH, Kaufingerstraße 24, 80331 München, Germany

Printed and bound by CPI Group (UK) Ltd, Croydon, CR0 4YY
11/09/2026
02215124-0008